Real Options "in" Projects and Systems Design

Tao Wang

Real Options "in" Projects and Systems Design

Identification of Options and Solution for Path Dependency

VDM Verlag Dr. Müller

Impressum/Imprint (nur für Deutschland/ only for Germany)

Bibliografische Information der Deutschen Nationalbibliothek: Die Deutsche Nationalbibliothek verzeichnet diese Publikation in der Deutschen Nationalbibliografie; detaillierte bibliografische Daten sind im Internet über http://dnb.d-nb.de abrufbar.

Alle in diesem Buch genannten Marken und Produktnamen unterliegen warenzeichen-, marken- oder patentrechtlichem Schutz bzw. sind Warenzeichen oder eingetragene Warenzeichen der jeweiligen Inhaber. Die Wiedergabe von Marken, Produktnamen, Gebrauchsnamen, Handelsnamen, Warenbezeichnungen u.s.w. in diesem Werk berechtigt auch ohne besondere Kennzeichnung nicht zu der Annahme, dass solche Namen im Sinne der Warenzeichen- und Markenschutzgesetzgebung als frei zu betrachten wären und daher von jedermann benutzt werden dürften.

Coverbild: www.purestockx.com

Verlag: VDM Verlag Dr. Müller Aktiengesellschaft & Co. KG
Dudweiler Landstr. 99, 66123 Saarbrücken, Deutschland
Telefon +49 681 9100-698, Telefax +49 681 9100-988, Email: info@vdm-verlag.de
Zugl.: Boston, MIT, Diss., 2005

Herstellung in Deutschland:
Schaltungsdienst Lange o.H.G., Berlin
Books on Demand GmbH, Norderstedt
Reha GmbH, Saarbrücken
Amazon Distribution GmbH, Leipzig
ISBN: 978-3-639-10371-7

Imprint (only for USA, GB)

Bibliographic information published by the Deutsche Nationalbibliothek: The Deutsche Nationalbibliothek lists this publication in the Deutsche Nationalbibliografie; detailed bibliographic data are available in the Internet at http://dnb.d-nb.de.

Any brand names and product names mentioned in this book are subject to trademark, brand or patent protection and are trademarks or registered trademarks of their respective holders. The use of brand names, product names, common names, trade names, product descriptions etc. even without a particular marking in this works is in no way to be construed to mean that such names may be regarded as unrestricted in respect of trademark and brand protection legislation and could thus be used by anyone.

Cover image: www.purestockx.com

Publisher:
VDM Verlag Dr. Müller Aktiengesellschaft & Co. KG
Dudweiler Landstr. 99, 66123 Saarbrücken, Germany
Phone +49 681 9100-698, Fax +49 681 9100-988, Email: info@vdm-publishing.com

Printed in the U.S.A.
Printed in the U.K. by (see last page)
ISBN: 978-3-639-10371-7

Abstract

This research develops a comprehensive approach to identify and deal with real options "in" projects, that is, those real options (flexibility) that are integral parts of the technical design. It represents a first attempt to specify analytically the design parameters that provide good opportunities for flexibility for any specific engineering system.

It proposes a two-stage integrated process: options identification followed by options analysis. Options identification includes a screening and a simulation model. Options analysis develops a stochastic mixed-integer programming model to value options. This approach decreases the complexity and size of the models at each stage and thus permits efficient computation even though traditionally fixed design parameters are allowed to vary stochastically.

The options identification stage discovers the design elements most likely to provide worthwhile flexibility. As there are often too many possible options for systems designers to consider, they need a way to identify the most valuable options for further consideration, that is, a screening model. This is a simplified, conceptual, low-fidelity model for the system that conceptualizes its most important issues. As it can be easily run many times, it is used to test extensively designs under dynamic conditions for robustness and reliability; and to validate and improve the details of the preliminary design and set of possible options.

The options valuation stage uses stochastic mixed integer programming to analyze how preliminary designs identified by the options identification stage should evolve over time as uncertainties get resolved. Complex interdependencies among options are specified in the constraints. This formulation enables designers to analyze complex and problem-specific interdependencies that have been beyond the reach of standard tools for options analysis, to develop explicit plans for the execution of projects according to the contingencies that arise.

The framework developed is generally applicable to engineering systems. The book explores two cases in river basin development and satellite communications. The framework successfully attacks these cases, and shows significant value of real options "in" projects, in the form of increased expected net benefit and/or lowered downside risk.

Acknowledgements

My deepest thanks to Professor de Neufville for his guidance and help. He leads me to the cutting edge of the research of real options "in" projects and gives me critical advice. Besides helping me embark on the promising area of real options, he gives me a lot of important intangible assets, such as rigor and strictness.

My heartful thanks to my friends. They help me gather important information, teach me many real world practice and knowledge, and relax me during the strenuous work. I do not know how I can fully thank all the friends and reward them.

My sincere thanks to my family. My parents have supported me unwaveringly during my study in the US. As the only child of the family and when my mother was sick, all their understanding and encouragement helped me overcome various difficulties. I hope to deliver this fruit to my parents to thank them for all their nurture and care!

Table of Contents:

List of Figures:

List of Tables:

Notations:

Symbol	Meaning
a	Factor for economies of scale for satellite system
$AMAX_s$	Maximum head at site s
$AMIN_s$	Minimum head at site s
A_{st}	Head at site s for season t
A_{sty}	Head at site s for season t year y
b	Factor for economies of scale for satellite system
c	Value of a call option
$CAPD_s$	Maximum feasible storage capacity at site s
$CAPP_s$	Upper bound for power plant capacity at site s
Cap_s	Incremental capacity a evolutionary architecture design s adds over the previous design
crf	Capital recovery factor
C_s	Cost coefficient for satellite architecture design s
$\overline{c}_{st}$	Part of flow at site s season t to be used in the construction period to ensure a full reservoir of the next period
c_j	Cost coefficients for design parameters
d	Down factor
D_A	Satellite antenna diameter in meter
D_i^q	Demand for satellite system for stage i under scenario q
E	Economic coefficients on engineering systems
e	Economic coefficients on engineering systems
e_p	Power factor
e_s	Power plant efficiency at site s
E_{st}	Average flow entering site s for season t
E_{sty}	Average flow entering site s for season t year y

f	The ratio of average yearly power production during the construction period over the normal production level, or option price
FC_s	Fixed cost for reservoir at site s
ΔF_{st}	Increment to flow between sites s and the next site for season t
h	Satellite orbital altitude in Kilometer
i	Stage index
H_s	Capacity of power plant at site s
h_t	Number of hours in a season
ISL	The use of inter satellite links, a binary variable in 0 or 1.
j	Design parameter index
k	Node index
K	The exercise price of an option
k_t	Number of seconds in a season
l	Node index
LCC	Discounted life circle cost
m	In a binomial tree, the mth nodes of a stage
n	Configuration index
p	Value of a put option, or risk-neutral probability
$P(k)$	The path from the root node 0 at the first stage to a node k
P_{ist}^q	Hydroelectric power produced at site s for season t for time period i for scenario q
p^q	The probability for a scenario q
P_{st}	Hydroelectric power produced at site s for season t
P_{sty}	Hydroelectric power produced at site s for season t year y
P_t	Satellite transmit power in Watt
PVC_i	Factor to bring cost in the ith period back to year 0
PV_i	Factor to bring 10-year annuity of benefit back to the present value as of year 0 (now)

PVO_i	Factor to bring the annuity after the 3 10-year periods till the end of 60 year life span of a project
q	Scenario index
Q	The number of terminal nodes
$Q_{in,t}$	Upstream inflow for season t
r	Discount rate (in some cases, specifically risk free interest rate)
RMB	China currency unit
$\mathbf{R^q}$	The real options decision variables corresponding to scenario q
R^q_{ij}	The decision on whether to build the feature according to jth design parameter for ith time stage in scenario q
s	Project index (site index for water resource case and architecture design index for satellite system case)
S	Stock price
S^q_i	Value of underlying asset at time stage i in scenario q
S_{st}	Storage at site s for season t
S_{sty}	Storage at site s for season t year y
t	Season index
$\mathbf{t}$	Technical limits on engineering systems
T	Time to expiration
$\mathbf{T}$	Technical coefficients on engineering systems
ΔT	Time interval between two consecutive stages
T_{st}	Target release at site s for season t
u	Up factor
VC_s	Variable cost for reservoir at site s
V_S	Capacity of reservoir at site s
$\mathbf{X}$	Design vector for satellite architectural design decisions
X^q_{ist}	Average flow from site s for season t for time period i for scenario q
X_{st}	Average flow from site s for season t

X_{sty}	Average flow from site s for season t year y
y	Year index for the simulation model
Y_j	Design parameter
$\overline{\mathbf{Y}}$	The set of satisfactory configurations of design parameters
$\overline{Y_j}$	Design parameter in most satisfactory design
yr_s	Integer variable indicating whether or not the reservoir is constructed at site s
Y_{st}	Power factor at site s for season t
z	Configuration decision variable, only one configuration can be built
β_j	Benefit coefficients for design parameters
β^p	Hydropower benefit coefficient
β_{Pi}^q	Hydropower benefit coefficient for period i and scenario q
δ_i	the set of nodes corresponding to time stage i
δs	Variable cost for power plant at site s
ω^q	A joint realization of the problem parameters
ω_i^q	The uncertain parameters for time stage i in scenario q
φ	Real options constraints
ε	Satellite minimum elevation angle in degrees
σ	Volatility
μ	Drift rate
Π	Outcome, or value of portfolio

Chapter 1 Introduction

Forecasts are "always wrong"!? That is, comparisons made between forecasts and the actual realizations one or two decades later show wide discrepancies. Differences of a factor of two, up or down, are typical. Moreover, the demand that does occur is frequently substantially different from what was predicted. Only exceptionally do long-term forecasts actually hit the mark, or actual demand exceeds expectations.

- In 1991, the forecasts for the US terrestrial cellular phone market were considered optimistic. It expected up to 40 million subscribers by the year 2000 (Ciesluk et al., 1992). The standardization of terrestrial cellular networks resulted in over 110 million subscribers in 2000, 275% of the projection.
- In 1991, forecasts for the satellite cellular phone market are similar to that of the terrestrial cellular phone market. Initiatives like Iridium and Globalstar were encouraged by the absence of common terrestrial cellular phone standards and slow development of cellular networks at that time. Iridium was designed according to the forecast of 3 million subscribers. It only aroused the interest of 50K initial subscribers and filed for bankruptcy in August 1999. Globalstar went bankrupt on February 2002.
- "Heavier-than-air flying machines are impossible." (Lord Kelvin, British mathematician, physicist, and President of the British Royal Society, c. 1895)
- "A severe depression like that of 1920-1921 is outside of the range of possibility." (Harvard Economic Society – Weekly Letter, November 16, 1929)

As a general rule, we can expect that the actual long-term future will be different from what was projected as the most likely scenario. Designers must expect that the engineering system will have to serve any one of a range of possibilities, and manage uncertainties proactively.

1.1. The problem

With the recognition that design context is uncertain, how to manage uncertainties proactively? Designing flexibility into engineering systems with real options! Real options theory is a formal way to define flexibility. It models flexibility like financial options and uses options theory to value flexibility quantitatively, and thus enable us to compare the value of flexibility with the cost of acquiring such flexibility to decide if we want to design the flexibility into the systems.

Since the options theory has not yet been extended to physical design as our literature research suggests, several problems have to be solved before real options can be applied in designing flexibility into physical systems. The most important two are:

- Where is the flexibility? Unlike financial options that are well-defined legal contracts, real options in physical systems need to be defined before analysis. The problem is not that real options "in" projects are impossible to find. The difficulty lies in that there are too many variables, and thus too many possible real options; however, less than a small fraction of the possible real options can be considered. The designers need to identify the "best" opportunities, the real options most likely to offer good flexibility in the uncertain environment.
- How to value highly interdependent/path-dependent/complex options? In physical systems, we meet many kinds of technical dependency that are not present in the finance world. Large-scale engineering systems feature a great number of technical constraints. These constraints will force real options "in" them to present highly interdependent and path-dependent characteristics that are not typical in financial options. For example, the water flow continuity specifies how the power generation capacity of a downstream dam will change when an upstream dam is built. The standard financial options valuation techniques such as Black-Scholes formula or binomial tree are impotent before such complex and

special interdependent/path-dependent options. We need an innovative options valuation approach to deal with technical interdependency/path-dependency.

-

1.2. Research opportunity

This work studies engineering design and real options from a unique perspective - engineering systems. Engineering Systems Division (ESD) is a new organization at MIT tackling the large-scale engineering challenges. The nature of ESD is interdisciplinary to study the common traits across a wide range of engineering, for example, aeronautics, astronautics, civil and environmental engineering, electrical and computer engineering, etc.

This study integrates three threads of research – engineering systems with regard to uncertainties, real options, and mathematical programming – to attack an area currently known little by systems designers. The area is proactive management of uncertainties in engineering systems. See Figure 1-1.

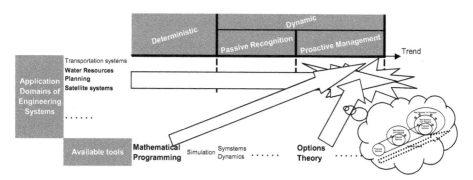

Details see Figure 1-2

Figure 1-1 Overall Picture of Research

The first thread - engineering systems with regard to uncertainties

The third thread is the major trend that engineering systems design develops from deterministic to dynamic. See the horizontal arrow in Figure 1-1.

By deterministic, we mean the design practice based on design parameters that are not sequences of probability functions at multiple points in time. A typical deterministic design practice forecasts expected values of uncertain parameters, and uses those expected values as inputs for further analysis and design. Optimization with such design parameters often leads to economies of scale. To a first order, economies of scale are common in industries where cost largely depends on the envelope to the structure (a quantity expressed in terms of the square of the linear measure), and capacity depends on the volume (a quantity expressed in terms of the cube of the linear measure). In such situations, total cost grows approximately to the 2/3 power of capacity (de Neufville, 1990; Chenery, 1952).

By dynamic, we mean the design practice taking into consideration the design parameters that are sequences of probability functions at multiple points in time. For example, in the design of a series of hydropower stations on a river, the dynamic nature of water flow is considered. The water flow follows a certain distribution and is taken as a family of random variables indexed by time. Using the dynamic thinking of design, economies of scale resulted from deterministic design may disappear because of the possibility of underutilization of the facility when the circumstances are unfavorable. Dynamic thinking has a more accurate understanding of reality than deterministic thinking, and thus provides better designs in a forever-changing world.

In practice, dynamic thinking is sometimes only applied to the technical part of the design, while the economic or social perspectives are dealt with deterministically. For example, people may design a series of hydropower stations on a river with the understanding that water flow is stochastic, but take another important uncertain variable, price of electricity, as constant using the forecasted expected value. This kind of incomplete dynamic thinking prevails if designers are solely focused on the engineering side while losing sight of important economic and social uncertainties. The incomplete dynamic thinking without sufficient attention to both technical and social stochasiticity will render inadequate, if not misleading, results.

Moreover, in the reign of dynamic design, we are developing from passive recognition to proactive management of uncertainties. Passive recognition of uncertainty leads people to *do* something *now* to withstand the worst scenarios; proactive management of uncertainty leads people to **prepare** to do something **in the future** to avoid downside risks and take advantage of upside potential. In some sense, traditional robustness is a way of passive recognition of uncertainty[1]. Robustness is a property to withstand unfavorable situations without actively changing a system. Proactive management of uncertainty, however, requires the design can be reinforced in a timely manner when unfavorable things happen, not necessary to do everything now. The non-traditional part of proactive management of uncertainty is about taking advantage of upside potential. We can be opportunist to benefit from a better-than-expected situation! In short, passive recognition of uncertainty misses the part of reality that people can take actions when reality unfolds.

Since engineering systems cover a broad range, the book will study in-depth a specific area of engineering systems – water resources systems – and develop a general framework applicable to other systems, with a brief case example on satellite communications systems to show the generalizability of the framework.

---

[1] Here we do not intend to devalue the importance of robustness. In many cases, robustness is a must in design.

The second thread – options theory

After the modern options theory was founded by Black, Scholes, and Merton (1973)[1], options theory and thinking has been gradually extended to broader areas in both finance and non-finance, from financial options to real options "on" projects to real options "in" projects. See the rightmost arrow in Figure 1-1. Options theory yields insights into uncertainty and flexibility that enhance the ability of human beings to deal with forever-changing environment.

A real option is a right, but not obligation, to do something for a certain cost within or at a specific period of time. The valuation of real options provides important insights into the value of opportunity or flexibility. Real options are extension of financial options. While financial options are traded on exchanges and over-the-counter markets, real options are not traded. Real options is more a methodology for valuing investments or designing flexibility.

Real options can be categorized as those that are either "on" or "in" projects (de Neufville, 2002). Real options "on" projects are financial options taken on technical things, treating technology itself as a 'black box'". Real options "in" projects are options created by changing the actual design of the technical system. For example, de Weck et al (2004) evaluated real options "in" satellite communications systems and determined that their use could increase the value of satellite communications systems by 25% or more. These options involve additional fuel for orbital maneuvering system (OMS) onborad satellites in order to achieve a flexible design that can adjust capacity according to need.

One dimension of the general development of options theory is depicted in Figure 1-2. With the development of the options theory (as the arrow represents passage of time), the scope of application (as represented by the area of bubbles) is expanding, from financial options to real options "on" projects to real options "in" projects. Real options

---

[1] Their landmark Black-Scholes model (1973) won the Nobel Prize in 1997

"in" projects further expand the options thinking into physical systems, adding flexibility into physical systems systematically with full awareness of uncertainties. With the success of the options theory and its key insights into uncertainty, it has bright prospects to improve engineering systems design in meeting customer demands and regulatory requirements as well as increasing its economical feasibility or profitability.

In general, real options "in" systems require a deep understanding of technology. Because such knowledge is not readily available among current options analysts, there have so far been few analyses of real options "in" projects, despite the important opportunities available in this field. Moreover, the data available for real options "in" project analysis is of much poorer quality than that of financial options or real options "on" projects. Real options "in" projects are different and need an appropriate analysis framework - existing options theory has to adapt to the new needs of real options "in" projects.

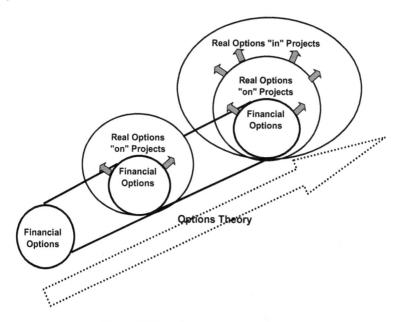

Figure 1-2 Development of Options Theory

The third thread - mathematical programming

Mathematical programming seeks to maximize or minimize a real or integer function given certain constraints on the variables. Mathematical programming studies mathematical properties of maximizing or minimizing problems, formulates real world problems using mathematical terms, develops and implements algorithms to solve problems. Sometimes mathematical programming is mentioned as optimization or operations research. Mathematical programming has many topics. Some of the major topics are linear programming, non-linear programming, and dynamic programming. Mathematical programming has also developed from deterministic to dynamic, and stochastic versions of the topics have been developing.

Mixed-integer programming is a useful tool for the purpose of options analysis. A 0-1 binary variable can neatly represent the options decision regarding whether to excise an option or not. Stochastic programming is the method for modeling optimization problems that involve uncertainty. The goal of the formulation is to find some policy that is feasible for the data instances and maximizes the expectation of some function of the decisions and the random variables. Combining stochastic programming and mixed-integer programming is a promising tool to model real options.

The Intersection of the three threads

Overall, this study identifies a point where engineering systems design needs proactive management of uncertainties, real options "in" projects need more development, and stochastic mixed-integer programming provides an appropriate tool for the study (see Figure 1-1). The intersection of several areas is often fertile soil to integrate existing knowledge and generate innovative ideas inspired by the knowledge of distinctive areas.

General applicability and case examples

As shown in Figure 1-3, the contribution of the work would be providing a **general** approach that large-scale engineering systems can use for their specific systems to design flexibility. Case examples will be drawn on two different large-scale engineering systems, namely, water resources systems and satellite communications systems, showing the generalizability of the approach as well as helping illustration of the approach. The case examples also provide a foundation for water resources engineers and satellite systems engineers to add more details and serve for actual application. The core of the approach is a screening model to identify options and a mixed-integer programming real options timing model to analyze options.

Analyis with Options Thinking		Engineering Systems		
		River Basin Development	Satellite Systems	...
	General	*This work* —		
	More Details	▼		

Figure 1-3 Contribution and generalizability of the work

1.3. Theme and Structure of the Book

This book proposes a two-stage options identification and analysis framework to design flexibility (real options) into physical systems, with case examples on river basin development and satellite communications systems.

The structure of the book is as in Figure 1-4: Chapter 2 reviews literature on engineering systems, water resources planning, mathematical programming, and options theory. Chapter 3 introduces standard options theory. Its focus is on the options valuation models and their key assumptions. Chapter 4 introduces real options, especially the difference between real options "on" and "in" projects. Chapter 5 develops the general two-stage framework for options identification and analysis in engineering systems. Chapter 6 presents a detailed case example on a river basin using the two-stage framework. Chapter 7 develops some policy implications for the framework on designing flexibility into physical systems. Chapter 8 summarizes and concludes the book.

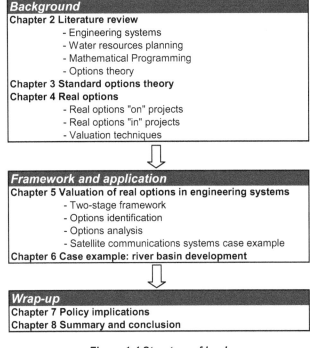

Figure 1-4 Structure of book

Chapter 2 Literature Review

This study integrates three threads of research – engineering systems with regard to uncertainties, real options and mathematical programming – to attack an area currently known little by systems designers. The area is proactive management of uncertainties in systems design. This chapter explores historical and intellectual developments in areas related to the study in this book.

2.1. Engineering Systems and uncertainties

Engineering systems is a new discipline spawned by the development of modern engineering science. As Roos [1998] noted:

> "Engineering systems are increasing in size, scope, and complexity as a result of globalization, new technological capabilities, rising consumer expectations, and increasing social requirements. Engineering systems present difficult design problems and require different problem solving frameworks than those of the traditional engineering science paradigm: in particular, a more integrative approach in which engineering systems professionals view technological systems as part of a larger whole. Though engineering systems are very varied, they often display similar behaviors. New approaches, frameworks, and theories need to be developed to understand better engineering systems behavior and design."

In comparison to Engineering Systems, the other current paradigm in academic settings is referred to as Engineering Science. We can study the development of Engineering Science and Engineering Systems on a time line. The invention of computers in the 1940s' and the following development of computers push the development of engineering studies. In one direction, it pushes the rapid development of Engineering Science on every area in a more and more detailed and thorough fashion. In the other direction, starting with the development of system models – use of linear programming in major engineering areas, the discipline of Engineering Systems grows gradually. In the 1970s', the engineering education and research extends to the intersection of social factors and engineering. Carnegie Mellon University established its Engineering and Public Policy (EPP) program, and Prof Richard de Neufville founded the Technology and Policy Program (TPP) at MIT. Increasingly, engineering research and practice are overlapping more and more with each other, and with social sciences. In 2000, MIT started its experiment on Engineering Systems Division (ESD) to address these interdisciplinary needs and the formation of the new disciplinary area of Engineering Systems. For the MIT ESD history, Roos [2004] had a good introduction. Hastings [2004] described the plan for the future of MIT ESD. He said, "What is needed is the development of a holistic view of these systems that takes into account all the issues associated with them." Figure 2-1 depicts the time line. In short, Engineering Systems offer large strategic view, while Engineering Science gives a specific technical view.

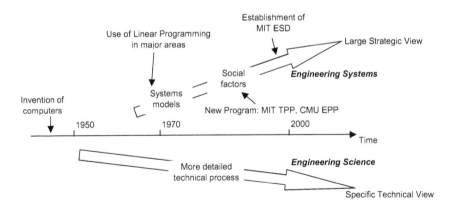

Figure 2-1 Development of Engineering Systems

While engineering systems face a lot of issues, two of the issues are most important: uncertainty and complexity. The other key aspects of engineering systems are technical, people/organization, and context. See Figure 2-2. Moses [2004] discussed fundamental issues associated with large-scale, complex engineering systems, such as complexity and uncertainty, flexibility. This book focused on one of the two important aspects of uncertainty. De Neufville et al. [2004] discussed the particularly significant long-term fundamental issue for planning, design and management of engineering systems - uncertainty. Moses [2004] wrote "non-traditional system properties are of great interest in Engineering Systems, partly because some of them, such as flexibility and sustainability, have not been sufficiently studied."

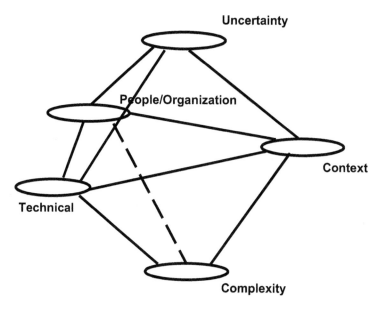

Figure 2-2 Three Dimensions of Engineering Systems Linked to Each Other and to Two High Impact Aspects of All Such Systems [Source: MIT ESD Symposium Committee, 2002]

"Engineering systems designers have long recognized that design context is uncertain and forecasts are 'always wrong,' as numerous observers have documented (such as Ascher, 1978; Makridakis, 1979a, 1979b, 1984, 1987, 1989, 1990; US Office of Technology Assessment, 1982; de Neufville and Odoni, 2003). That is, comparisons made between forecasts of the level of demand and the actual realizations one or two decades later show wide discrepancies. Differences of a factor of two, up or down, are typical. Moreover, the demand that does occur is frequently substantially different from what was predicted. For example, the forecasts of the need for telephone services made in the 1980s failed to anticipate the explosion in use (and need for infrastructure) caused by the use of cell phones, and did not anticipate the change in the distribution of connection times – longer as regards constantly on Internet connections, and shorter as regards text messaging. Only exceptionally do long-term forecasts actually hit the mark. As a general rule, we can expect that the actual long-term future demand for engineering systems design will be different from what was projected as the most likely scenario. Designers must expect that the engineering system will have to serve any one of a range of possibilities, and manage uncertainties proactively." (de Neufville, Ramírez, and Wang, 2005).

Though people understood that the design context is uncertain, people still applied deterministic tools to analyze the problem because the computational cost was not affordable during the early years. With the development of computers and theories, the trend of design enters into a dynamic reign. People study the uncertain context and recognize the inherent uncertain world. Furthermore, people are exploring how to manage the uncertainties proactively with smarter design.

2.2. Water Resource Planning

Engineering systems study covers a lot of different systems, for example transportation systems, satellite communication systems, and water resources systems. One of the systems focused in this book is water resource planning, especially with regard to facilities design, an area that came into maturity by 1980, though the prevailing methodology does not consider the design issues in the full context of the changing and uncertain world. This book develops an analysis framework that in some way builds on the standard water resources planning, and extends the usage of the framework to other engineering systems with a case example in satellite communications.

2.2.1. Historical development

After economic analysis was brought into water resource planning studies, research on water resource planning with regard to facilities design can be divided into three phases: mathematical programming, multiobjective analysis, and risk recognition.

Mathematical programming

Mathematical programming is the major tool used in water resource planning. Maass et al. (1962) summarized the contributions of the Harvard Water Program (1955 – 1960). They introduced the most advanced techniques at that time: such as linear programming, mathematical synthesis of streamflow sequences, and computer simulation. This study laid foundations for future development of water resource planning.

Hufschmidt and Fiering (1966) described a river basin computer simulation model thoroughly. Before computers became generally available, the simulation models were physical models scaled to maintain static or dynamic similitude. The simulation model of Hufschmidt and Fiering dealt with a large number of randomly selected designs, included generation of long hydrologic sequences, and measured outcomes in economic terms. Their model was much more advanced than the then prevailing physical simulations. They tested 20 randomly selected designs for 250 years (3000 months) of simulated operations. The 3 designs with the highest net benefits were subjected to further analysis using single-factor and marginal analysis methods. The model was written in FORTRAN. The computer used was the IBM 7094 that had 32,768 directly addressable memory (whereas modern computers have memory measured in Gigabytes!)   It took about 7.5 minutes of computation for one single operation of simulation for 250 years.

Jacoby and Loucks (1974) developed an approach to the analysis of complex water resource systems using both optimization and simulation, not as competing techniques but as interacting and supplementing methods. They used preliminary screening models to select several alternative design configurations; then they simulated the preferred designs using the same annual benefit, loss, and cost functions. They estimated the expected benefits for each state for each 5-year period from 1970 to 2010.

Major and Lenton (1979) edited a book on a study, led by David Marks, of the Rio Colorado river basin development in Argentina.   Their system of models consisted of a screening model, a simulation model and a sequencing model. The screening model is a mixed-integer programming model with about 900 decision variables, including 8 0-1 integer variables, and about 600 constraints. Objectives were incorporated either into the objective function or as constraints on the system.   Multiobjective criteria underlay the whole formation of the model. The simulation model evaluated the most promising configurations from the screening model in terms of net benefits and hydrological reliability. The runs from the simulation model improved the configurations from the screening model. The simulation model was operated with 50 years of seasonal (4-month) flows. The sequencing model scheduled a candidate configuration optimally in 4 future time periods, taking into account benefits over time, budget constraints, constraints on the number of farmers available, and project interrelationships. The mixed-integer programming sequencing model had about 60 continuous variables, 120 integer variables, and 110 constraints depending on the exact configuration being modeled.

Loucks, Stedinger, and Haith (1981) summarized the art of water resource planning until then, such as evaluation of time streams of benefits and costs, plan formulation, objective functions and constraint equations, Lagrange multipliers, Dynamic Programming, Linear Programming, Simulation, probability and distribution of random events, stochastic processes, and planning under uncertainty.

After 1980, there are relatively fewer articles on the topic regarding facilities design and planning of big river basins. The interests shift more to operating policies of reservoirs, water supply, water quality, environmental issues. This trend is coherent with the trend in the Western countries to stop building or even tear down dams, in favor of more environmental considerations rather than short term gain on agricultural and hydropower benefits of big water projects. It is also partly because the art had been pretty mature that developments in water resource planning theory became less concerned with the design of water resource facilities. Another reason for this is that most economical sites for water projects have almost all been built in North America or Europe, the remaining sites are marginal and economical benefits can not overweigh the environmental costs to society. With the decrease of interest in the construction of big water river projects, the literature on this topic has been less, correspondingly. William (1996) argued two basic approaches of water resources planning – that of the Corps of Engineers and other construction agencies and that of the U.S. Environmental Protection Agency (EPA) and other regulatory agencies - are both incomplete. The requirements of the various regulatory approaches are making it almost impossible to construct major facilities for any purpose, and water resource analysts were reluctant to challenge them. A more complete approach is needed to reach better results.

The development of big dams in developing countries has not slowed down. Sinha, Rao and Lall (1999) presented a screening model for selecting and sizing potential reservoirs and hydroplants on a river basin. A linked simulation-optimization framework is used. The objective function is to meet annual irrigation and hydropower demands at prescribed levels of reliability. Sizing of reservoirs and hydroplants, and evaluation of objective function and constraints and their derivatives are done as part of simulation. The formulation is applied to river basins in India. Sinha, Rao, and Bischof (1999) presented an optimization model for selecting and sizing potential reservoir and hydropower plant sites on river basins. The model used a behavior analysis algorithm that allows operation of the reservoir system with realistic operating policies. The model is developed in the context of river basins in India. Dahe and Srivastava (2002) extended the basic yield model and presented a multiple-yield model for a multiple-reservoir system consisting of single-purpose and multipurpose reservoirs. They argued the model could act as a better screening tool in planning by improving the efficiency and accuracy of detailed analysis methods such as simulation. The model is applied to a system of eight reservoirs in India.

The application and development of simulation-optimization frameworks in the US has never been stopped, though it is less on construction and more on management of existing facilities. Nishikiwa (1998) used a similar simulation-optimization approach for the optimal management of the city of Santa Barbara's water resources during a drought. The objective is to minimize the cost of water supply subject to water demand constraints, hydraulic head constraints to control seawater intrusion, and water capacity constraints. The decision variables are monthly water deliveries from surface water and ground water. Draper et al. (2003) presented an economic-engineering optimization model of California's major water supply system. They argued that the economic-engineering optimization model could suggest a variety of promising approaches for managing large systems. The model is deterministic. Lefkoff and Kendall (1996) evaluated yields from a proposed ground-water storage facility using a nonlinear optimization model of the California State Water Project (SWP) and the Central Valley Project (CVP). Model constraints include the major hydrologic, regulatory, and operational features of both projects, including mass continuity, facility capacities, regulatory standards, and delivery contracts.

Development of powerful computers and computer programs such as GAMS improved the performance of previous methodologies and enabled the solution of much bigger problems. Meanwhile, researchers have never stopped to search for better algorithms for water resources optimization problems. Anderson and Al-Jamal (1995) used non-linear programming to develop a methodology for simplification of complicated hydraulic networks. Kim and Palmer (1997) presented a Bayesian Stochastic Dynamic Programming model to investigate the value of seasonal flow forecasts in hydropower generation. Watkins and McKinney (1997) introduced Robust Optimization (RO) as a framework for evaluating these trade-offs and controlling the effects of uncertainty in water resources screening models. Sinha and Bischof (1998) observed that the automatic differentiation method may greatly benefit the convergence of the optimization algorithm of reservoir systems to determine optimal sizes for reservoirs.

Multiobjective analysis

There are many different objectives for water resource planning, such as national or regional income maximization, income redistribution, environmental quality, social well-being, national security, self-sufficiency, regional growth and stability, and preservation of natural areas. Some objectives can be easily expressed in monetary terms, while some cannot. Some (or all) objectives are conflicting and non-commensurable. Multiobjective analysis does not yield a single optimal solution, but identifies the production possibility frontier (or in other words, Pareto Frontior) and trade-offs among objectives.

Cohon and Marks (1974) discussed an application of multiobjective theory to the analysis of development of river basin systems. Major (1974) provided a case of the application of Multiobjective analysis to the redesign of the Big Walnut Dam and reservoir in Indiana. Major and Lenton (1979) demonstrated the application of mathematical modeling and multiobjective investment criteria to river basin development.

Risk recognition

The above-mentioned studies transformed technical parameters into expected total annual net efficiency benefits (or the utility for human and society) and maximized the net efficiency benefits (or utility) to obtain the "optimal design". They carefully considered technical uncertainties, such as that of waterflow, and used dynamic models. However, they did not take into account uncertainties in the human and social sphere. Ignoring the human and social uncertainties, the methodology cannot reach the "optimal design" (if such designs exist) by simply recognizing the technical uncertainties. Any technical systems are to serve human's needs.

Recent studies on water resource planning are more explicitly taking human, social, environmental uncertainties into account. Morimoto and Hope (2001, 2002) applied probabilistic cost-benefit analysis to hydroelectric projects in Sri Lanka, Malaysia, Nepal, and Turkey. Their results were in the form of distributions of NPV. These studies recognized the uncertainties from human, social, and environmental perspectives. But they did not take into account the value of options, or the flexibility. Decision-makers do not passively succumb to fate and they will respond to the uncertain environment. Increasingly, the real options concept has been introduced to water resources planning and management (US National Research Council, 2004).

2.2.2. Other important references

Based on Manne (1961, 1967), Hreisson (1990) dealt with the problem of obtaining the "optimal design" of hydroelectric power systems regarding sizing and sequencing. The conclusion was to make the current marginal value of each new project equal to the discounted weighted average of the long-term marginal unit cost of all future projects. He further investigated economies of scale and optimal selection of hydroelectric projects. The tradeoff between large and small projects was studied by weighting the lost sales during the period of excess capacity against the benefit of using larger projects due to the economies of scale. All his studies were based on a deterministic view. If uncertainties regarding the demand and supply are high, the rules Hreisson developed may be misleading.

The fact that there is great uncertainty about the future loads on the infrastructure system has two implications for design, as Mittal (2004) has documented: the optimal size of the design should be smaller than that defined by Manne (1961, 1967), that is, planned for a shorter time horizon; yet the design should be easy to adjust for the range of possible long-term futures.

Aberdein (1994) illustrated the case of excess electricity on the South Africa interconnected grid resulted from the mismatch between planned capacity and actual demand. He stressed the importance of incorporating risks into power station investment decisions.

Cai, McKinney,and Lasdon (2003) argued that the interdisciplinary nature of water resources problems requires the integration of technical, economic, environmental, social, and legal aspects into a coherent analytical framework. Their paper presented the development of a new integrated hydrologic-agronomic-economic model, with the ability to reflect the interrelationships between essential hydrologic, agronomic, and economic components, and to explore both economic and environmental consequences of various policy choices.

Some papers available on the website of the World Commission on Dams (http://www.dams.org) are very helpful. For example, Fuggle and Smith (2000) prepared a report presenting background information on China's dam building program, financing, and policy development. Clarke (2000) reported the findings of a global dam survey covering 52 countries and 125 large dams. This report provides data for the uncertain factors, such as the project schedule performance data, actual-to-planned hydropower energy out, and many others.

2.3. Mathematical Programming

Mathematical programming seeks to maximize or minimize a real or integer function given certain constraints on the variables. Mathematical programming studies mathematical properties of maximizing or minimizing problems, formulates real world problems using mathematical term, develops and implements algorithms to solve the problems. Sometimes mathematical programming is mentioned as optimization or operations research. As early as 1665, Newton developed his method of finding a minimum solution of a function. The first book on mathematical programming – *Methods of Operations Research* (Morse and Kimball) - was written in 1946, but it was classified because of its extensive contribution in World War II - everything from how best to use radar or hunt submarines to getting supplies to the troops efficiently. In 1951, the unclassified version of *Methods of Operations Research* got published.

Fundamental mathematical programming topics

Mathematical programming has many topics. Some of the fundamental topics are linear programming (first developed by Dantzig in 1948), non-linear programming (first developed by Kuhn and Tucker in 1951), and dynamic programming (first developed by Bellman in 1957).

A Linear Program (LP) is a problem that can be expressed as follows (the so-called Standard Form):

$$\text{Max} \quad \mathbf{cx}$$
$$\text{S.t.} \quad \mathbf{Ax} \le \mathbf{b}$$
$$\mathbf{x} \ge \mathbf{0}$$

where $\mathbf{x}$ is the vector of variables to be solved for, $\mathbf{A}$ is a matrix of known coefficients, and $\mathbf{c}$ and $\mathbf{b}$ are vectors of known coefficients.

A Nonlinear Program (NLP) is a problem that can be put into the form

$$\text{Max} \quad F(\mathbf{x})$$
$$\text{S.t.} \quad g_i(\mathbf{x}) \le 0 \quad \text{for } i = 1, ..., n \qquad \text{where } n >= 0$$
$$h_j(\mathbf{x}) \ge 0 \quad \text{for } j = n+1, ..., m \qquad \text{where } m >= n$$

That is, there is one scalar-valued function F, of several variables, that we seek to maximize subject function F given the constraints. NLP is a difficult subject. The most important challenge is how to find the global optimum, particularly when $F(x)$ is non-convex.

Dynamic Programming (DP) is an algorithmic technique in which an optimization problem is solved by caching subproblem solutions rather than recomputing them. An intuitive explanation of DP is to think what to do best at the last stage, and deduce the best decisions at the second last stage, and back and back, so on and so on... finally reach the first stage and obtain the optimal decisions for the first stage.

The previous section on water resources planning has examined mathematical programming techniques used historically in water resources planning. It is the most important tool in water resources planning, especially regarding physical water facilities designs and planning.

Stochastic programming

Stochastic programming is the method for modeling optimization problems that involve uncertainty. It can be termed as optimization under uncertainty. In stochastic programming, some data are random, whereas various parts of the problem can be modeled as linear, non-linear, or dynamic programming. Stochastic programming has been developing fast with the contribution from and important application in operations research, economics, mathematics, probability, and statistics.

Sengupta (1972) published a book on methods and applications of stochastic programming in its various aspects at that time, for example, chance-constrained programming, two-stage programming under uncertainty, programming with recourse, reliability programming.

Archetti, DiPillo and Lucertini (1986) edited a collection of papers on stochastic programming, in two main research areas: stochastic modeling and simulation and stochastic programming. The papers cover algorithms and applications of stochastic programming. Notably, there is a paper on stochastic programming – the distribution problem (Rinnooy Kan, 1983).

Birge and Louveaux (1997) published a textbook on stochastic programming. It provides examples on modeling stochastic programs and describes how a stochastic model is formally built. It also covers mathematical properties, models, solution algorithms, approximation and sampling methods, and refers readers to the original papers for details.

Mixed-integer Programming

A mixed-integer programming (often simply called integer programming) problem is the same as the linear or non-linear problem except that some of the variables are restricted to take integer values while other variables are continuous. Mixed-integer programming is a powerful tool that can help formulating discrete optimization problems, e.g., our real options "in" projects problem. Meanwhile, the mixed-integer programming is a much more difficult problem than linear programming.

Bertsimas and Tsitsiklis (1997) have two chapters on mixed-integer programming. They introduce mixed-integer programming formulations of discrete optimization problems and provide a number of examples. They cover the major classes of integer programming algorithms, including exact methods (branch and bound, cutting planes, dynamic programming), approximation algorithms, and heuristic methods (local search and simulated annealing). They also introduce a duality theory for integer programming.

Mixed-integer programming has been widely used in the water resources planning. Besides the literature mentioned in the section for water resources planning, Srinivasan, Neelakantan, Narayan, and Nagarajukumar (1999) presented a mixed-integer programming model for the operation of a water supply reservoir during critical periods that incorporates important performance indicators, such as reliability, resilience, and vulnerability. Draper et al. (2004) introduced the new California Water Resources Simulation Model for the planning and management of the State Water Project and the federal Central Valley Project. It is not for the design of water facilities, though a lot of aspects of model are similar to those used for design of facilities. System description and operational constraints are specified using a new water resources engineering simulation language. A mixed integer linear programming solver efficiently routes water through the system network given the user-defined priorities or weights. Simulation cycles at different temporal scales allow for successive layering of constraints.

Stochastic mixed-integer programming

Stochastic mixed-integer programming is the most important tool this thesis suggests to deal with the path-dependency problem of real options valuation. The goal of the formulation is to find some policy that is feasible for the data instances and maximizes the expectation of some function of the decisions and the random variables.

With integer decision variables representing different possible decisions and a scenario tree approach to model uncertainty, a multi-stage stochastic integer programming can deal with stochastic optimization. Ahmed, King, and Parija (2003) used stochastic mixed-integer programming to formulate a capacity expansion problem, and outlined a branch and bound algorithm to solve the problem of capacity expansion under uncertainty.

According to Birge and Louveaux (1997), mathematical properties to be exploited are scarce for stochastic mixed-integer programming. There are few general efficient methods. Some techniques have been developed to deal with specific problems or particular properties, such as integer L-shaped method, simple integer recourse, binary first-stage variables, extensive forms and decomposition, and asymptotic analysis. The inadequacy in stochastic mixed-integer programming leads to the understanding that we may not be able to find a global optimum for our real options "in" projects problem, and we should be more realistically expecting improvement over current decision practice rather than reaching an optimal decision.

Software for mathematical programming

Because of the needs of mathematical programming application, software packages have been well developed. Two widely used are CPLEX and GAMS. As far as mixed-integer programming is concerned, GAMS offers a free trial version up to 300 discrete integer variables. Its full version is much more powerful. CPLEX is well acclaimed for its mathematical programming power. Its mixed-integer optimizer is deemed the best by many practitioners It employs a branch-and-bound technique that takes advantage of cutting-edge strategies and incorporates and expands on the latest results of worldwide research in mixed integer programming.

2.4. Options Theory

According to Hull (1999), stock options "were first traded on an organized exchange in 1973". The land-mark Black-Scholes model that won Nobel Prize in 1997 was initially developed for financial options in 1973 by Black, Scholes and Merton (Black and Scholes, 1973; Merton, 1973).

2.4.1. Development of real options method

Myers (1987) was one of the first to acknowledge that there are inherent limitations with standard discounted cash flow (DCF) approaches when it comes to valuing investments with significant operating or strategic options. He suggested that options pricing holds the best promise for valuing such investments.

Dixit and Pindyck (1994) stressed the important characteristic of irreversibility of most investment decisions, and the ongoing uncertainty of the environment in which these decisions are made. In so doing, they recognized the option value of waiting for better (but never complete) information. The focus of their book was on understanding investment behavior of firms, and developing the implications of this theory for industry dynamics and government policy.

Trigeorgis (1996) brought together previously scattered knowledge about real options. Comprehensively, he reviewed techniques of capital budgeting and detailed an approach based on the pricing of options that provides a means of quantifying flexibility. He also dealt with options interaction, the valuation of multiple options that are common in projects involving real options, and the valuation of the impact of competitive interactions. The methodology in this book was theoretical, and helped to shape more practical real options valuation techniques later on.

Besides theoretical developments, applications of real options are growing fast in business strategy, corporate finance, market valuation, contract valuation, security analysis, portfolio management, risk management, to engineering design. Real options methodology is applied in industries from natural resources development, real estate, R&D, information technologies, pharmaceutical, manufacturing, venture capital, government regulation, shipping, environmental pollution and global warming, to infrastructure.

Amram and Kulatilaka (1999) wrote an introductory book on real options, including financial options and applications of real options. But it does not provide a detailed practical methodology to evaluate real options. It gives readers an idea how widely real options can be applied.

The beginning of the twenty-first century sees a boom in the publication of books on real options with more focus on applications. Copeland and Antikarov (2001) wrote a controversial yet influential book on real options. Their major claim in the book is called ""Market Asset Disclaimer (MAD)", or "the present value of the cash flows of the project without flexibility (i.e., the traditional NPV) is the best unbiased estimate of the market value of the project were it traded." Based on MAD, they established a procedure to value real options that are commonly not market traded. Rogers (2002) describes his framework and insights in applying real options gained as a consultant with PricewaterhouseCoopers. The first half of the book brings together developments in strategy, real options, risk management and game theory to aid general managers. The second half of the book describes the tools and mathematical framework, in a broad range of areas, necessary to help the application of real options by technical analysts. Mun (2003) provides a qualitative and quantitative description of real options and multiple business cases and real-life scenarios. He explains practical uses of real options, minimizes coverage on replicating portfolios and focuses on risk neutral valuation. This is a good introduction book for MBA and other general users of Real Options. Brach (2003) explores how to apply real option valuation techniques on a regular basis from the view of a corporate practitioner. The author is an MD who has worked as a medical researcher for years. She has a first hand knowledge in health care and pharmaceutical research, and builds on this wealth of experience to develop many interesting examples of real options related to pharmaceutical drug development. Howell et al. (2001) offers a comprehensive book for starters on real options with advanced technical knowledge. The book tries to cover technical side with a transparent way to senior MBAs to advanced practitioners. Boer (2002) aims more at illuminating non-numerate readers of real options. He attempts to expose the reader in a nontechnical manner to the technique of real options without resorting to mathematical methods. The book is basically expository and attempts to build upon the readers' familiarity with financial options.

Although options thinking has been successfully applied to many areas, the application of real options in engineering has been slow, especially regarding building flexibility into the physical systems themselves. de Neufville (2002) suggested distinguishing between real options "on projects" and "in projects". The real options "on projects" concern the project but not system design, for example, the options imbedded in bidding for opening a mine.

The real options "in projects" concern the design elements of system and require detailed understanding of the system, for example, options for repositioning of constellations of communication satellites.

Over around 20-year development of the methodology, a great number of articles have been published on this topic. Another good information center of real options has been made by Marco Dias at http://www.puc-rio.br/marco.ind/main.html#contents. Borison (2003) offers an excellent overview on real options, and provides a critique of the major proposed analytic approaches for applying real options by their applicability, assumptions, and mechanics.

2.4.2. Application of real options

This book values options imbedded in river basin development and satellite communications systems. The literature review, thus, focuses more on the application of the valuation methodology, rather than theory.

Valuation approaches

Arbitrage-enforced pricing is a fundamental part of traditional options valuation. However, there are other ways to value options, which may be the only practical way if the no-arbitrage concept is irrelevant or cannot be used.

Arbitrage-enforced Real Options Valuation

As Baxter and Rennie (1996) explain, if there is arbitrage, it will enforce a price for the options. This price depends neither on the expected value nor on the particular distribution of the underlying assets. There are three categories of arbitrage-enforced valuation methods: the partial differential equation (PDE) such as the Black-Scholes formula, dynamic programming with binomial trees, and simulation.

The PDE method is standard and widely used in academic discussion because of the mathematical insights of the method. McDonald and Siegel (1986) studied the value of waiting to invest by PDEs. Siegel, Smith and Paddock (1987) valued offshore petroleum leases using PDE equations. Pindyck (1993) established PDEs to model project cost uncertainty, both technical and as regards input cost. Grenadier and Weiss (1997) studied the options pricing for investment in technological innovations.

The binomial tree method is based on a simple representation of the evolution of the underlying asset's value. It is a powerful yet flexible method to value real options. Cox, Ross, and Rubinstein (1979) developed this widely adopted method. Luenberger (1998) showed examples using binomial trees to value a real investment opportunity in a gold mine. Copeland and Antikarov (2001) elaborated how to use binomial trees to value real projects and proved this method, equivalent to PDE solution, is easy to use without losing the insights of the PDE model. Alternative lattice models can also be used to evaluate real options, such as the trinomial method used by Tseng and Zhao (2003). There are numerous variant approaches such as that of Copeland and Antikarov whose relative merit Borison (2003) discussed in detail.

On the other hand, these approaches assume path independence in the evolution of a system, that is, that the value of a design option does not depend on whether some other part of the system was or was not built. This assumption is often not correct, and requires a special analysis to overcome, as Wang and de Neufville (2004) have shown.

Simulation

With development of computer technology, big simulation programs can be constructed to value options that are very difficult to value by establishing/solving equations or building up binomial trees. Simulation can be either arbitrage-enforced or not. In the

1980s, Merck began to use simulation to value its R&D real options (Nichols, 1994). Tufano and Moel (2000) showed an elegant way to use Crystal Ball© to simulate the value of real options inherent in a bidding case for the Antamina Mine in Peru. Juan et al. (2002) suggested a simulation methodology to calculate multiple interacting American options on a harbor investment problem.

Expected value decision tree analysis

Decision Analysis (DA) has a long been used to value infrastructure investments (see for example Keeney and de Neufville, 1972). It accounts for the value of flexibility by structuring so that all uncertainties and possible future design decisions are explicitly considered. DA uses a decision tree to organize the sequence of events that occur and the contingent decisions. Once the tree has been laid out, DA finds the best possible decisions at each time for all the possible scenarios of events. The decision rule in DA is simple: choose the solution that offers the best expected value, a weighted average of the outcomes by their probability of occurrence (de Neufville, 1990).

In general, the calculation of the financial outcome of each scenario is based on NPV. Therefore, DA shares an important methodological weakness with NPV: it applies a fixed discount rate to the analysis instead of adjusting the discount rate to the level of risk associated with each scenario, as would be correct (Brealey and Myers, 2002). Strictly, decision tree analysis (DTA) is an expected value approach that does not yield a theoretically correct options value, unless the risk-adjusted discount rates and actual probabilities are used. To find risk-adjusted discount rate for each branch of the decision tree is difficult, if possible. However, people use DTA to illustrate the idea of real options and approximate the value of flexibility.

Faulkner (1996) showed how DTA could do "options thinking" valuation. Though this method does not provide a "correct" options value, it approximates the value, and more importantly, provides insights into options thinking.

Ramirez (2002) compared discounted cash flow methodology, decision tree analysis, and arbitrage-enforced real options. She examined their theoretical advantages and disadvantages, the assumptions, and information required. She also determined the consequences of the application of each approach on the nature of infrastructure projects. Although the real options methodology is theoretically superior in the pricing of flexibility, its implementation requires information usually not available for infrastructure assets. This makes the results of the analysis imprecise and complicates the process of identifying an optimal strategy.

Hybrid Model

Hybrid real options valuation combines the best features of decision tree analysis and real options analysis. Neely and de Neufville (2001) developed a hybrid real options valuation model for risky product development projects. The traditional valuation methods for risky product development are inadequate to recognize the value of flexibility while the real options method meets with difficulties to obtain the data necessary for a standard real options valuation. Their hybrid method analyzes market risks with real options analysis and project risks with decision tree analysis.

Summary

Table 2-1 illustrates five major types of valuation approaches with some examples. Figure 2-3 provides another way to categorize different methods in a 2-D space. Options analysis is the best both in terms of modeling of uncertainty and valuation of uncertainty. Lattice etc (including decision trees) is better than DCF in terms of modeling uncertainty, though both DCF and Lattice etc are inadequate with regard to valuation of uncertainty.

Borison (2003) described, contrasted, and critiqued the major proposed analytic approaches for applying real options. He observed relative strengths and weaknesses of the approaches and recommended which ones to use in what circumstance. He thought that the integrated approach (the hybrid model) provides the most accurate and consistent theoretical and empirical foundation, as does de Neufville and Neely (2001).

Table 2-1: Different Valuation Approaches with Examples

Examples	DTA	Hybrid Model	Arbitrage-enforced Real Options Valuation		
			PDE	Binomial Tree	Simulation
Automobile R&D Management (Neely and de Neufville, 2001)		X			
Bogota Water Supply Expansion (Ramirez, 2002)	X		X	X	
Merck (Nichols, 1994)			X		X
Kodak (Faulker, 1996)	X				

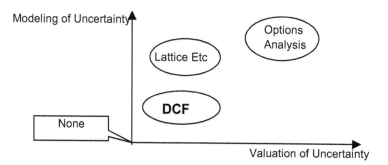

Figure 2-3 2-D View of Evaluation Practices [Source: Richard de Neufeille's class slides for MIT course ESD.71 developed in 2004]

Several important issues regarding real options valuation

<u>Underlying</u>

Financial options are based on underlying assets such as stocks, stock indices, foreign currencies, debt instruments, commodities, and futures contracts. They are traded in markets. Despite the fact that real options are not traded in markets, Mason and Merton (1985), and Kasanen and Trigeorgis (1993) maintained that real options may be valued similarly to financial options. The existence of a traded portfolio that has the same risk characteristics (i.e., is perfectly correlated) as a non-traded real asset is sufficient for real options valuation. Kulatilaka (1993) used the relative price of oil over gas to value the flexibility of a dual-fuel industrial steam boiler. Luenberger (1998) showed an example using the gold price as the underlying assets to value a real investment opportunity in a gold mine. Similarly, as shown in this thesis, it is possible to use energy price as the underlying asset to value a hydropower project under the assumption of a complete energy market.

However, in many cases, it is hard to find a priced portfolio whose cash payouts are perfectly correlated with those of the project, or in other words, to find market-priced underlying assets.

Is it possible to relax the definition of underlying to an agent that determines the value of a project, not necessarily market-traded? Copeland and Antikarov (2001) developed the assumption of "market asset disclaimer" and used the NPV of the underlying project as the underlying asset to build event trees to value real options. "Instead of searching in financial market," they recommended, "that you use the present value of the project itself, without flexibility, as the underlying risky asset—the twin security." (pp. 94) This method has the key disadvantages that it makes it impossible to identify the optimal strategy and blurs the exercise condition, because the NPV of a project is not readily market observable. How to find an appropriate underlying is an interesting question.

Volatility

Volatility is a measure of uncertainty, a key input of the options valuation. How to find the volatility is one of the key difficulties of application of real options if there is no market-traded underlying. Luehrman (1998) described three approaches: an educated guess by information such as industry or market as a whole, historical data such as record of investment returns and implied volatility of traded relevant options, and simulation of projected cash flow. He was among the first to present real options valuation techniques to a technically less advanced audience (*Harvard Business Review*) and appears to have a big impact on general managers. Copeland and Antikarov (2001) suggested, by first estimating the stochastic properties of variables that drive volatility, using Monte Carlo simulation to estimate it.

The estimate of volatility is often one of the weakest points of a real options valuation, since the valuation is usually sensitive to the volatility. Because volatility is distilled from a lot of information, it is practically impossible to estimate it for some real options valuation simply due to lack of data. Sometimes, therefore, the insights provided by a real options analysis are more important than a specific quantitative result.

Compound options and parallel options

Most real options are not well-defined simple options. They can be compound or parallel. They are often options on options (compound options) and the interactions between options are significant. For example, the opportunity to take a new product into mass production is an option on the R&D investment, whose value depends on the opportunity to proceed with R&D if the latter is exercised and successful. The methodology for valuing compound options is very important for the applicability of real options methodology in the real world. Parallel options are different options built on the same project, where those options interact. They are not necessarily mutually exclusive. For example, several possible applications of a new technology or several possible target markets of a new product. Oueslati (1999) described three parallel options for fuel cell development as automotive applications, stationary power, and portable power.

Geske (1979) developed approaches to the valuation of compound options. Trigeorgis (1993a and 1993b) focused on the nature of the interactions of real options. The combined value of a collection of options usually differs from the sum of their separate values. The incremental value of an additional option, in the presence of other options, is generally less than its value in isolation, and declines as more options are present. Oueslati (1999) explored the evaluation of compound and parallel real options in Ford's investment in fuel cell technology.

With all the developments in the application of the real options, the author is confident in applying the methodology on the river basin problem in this thesis. However, a lot of problems still await solutions. Without a market-observable underlying, the parameters used for the valuation are based on models that must be subjective, to a certain extent. So the model risks are not negligible in the method presented in this thesis.

Real Options applied in Energy and Natural Resources

The real options concept has been successfully applied in the energy industry. Siegel, Smith, and Paddock (1987) valued offshore petroleum leases using options, and provided empirical evidence that options values are better than actual DCF-based bids. Since then, research on real options on energy has been a hot topic. Miltersen (1997) presented methods to value natural resource investment with stochastic convenience yield and interest rates. Cortazar and Casassus (1997) suggested a compound option model for evaluating multistage natural resource investment. Cherian, Patel, and Khripko (2000) studied the optimal extraction of nonrenewable resources when costs accumulate.

Goldberg and Read (2000) found that a simple modification to the Black-Scholes model provides better estimates of prices for electricity options. Their modification combines the lognormal distribution with a spike distribution to describe the electricity dynamics. Bodily and Del Buono (2002) examined different models for electricity price dynamics, and proposed a new mean-reverting proportional volatility model. Dias (2002) gave a comprehensive overview of real options in petroleum.

Pindyck (1993) studied the uncertain cost of investment in nuclear power plants. He derived a decision rule for irreversible investments subject to technical uncertainty and input uncertainty. The rule is to invest if the expected cost of completing the project is below a critical number. The critical expected cost to completion depends on the type and level of uncertainty. Pindyck's work focused on finance issues of the project, the engineering model was not included in his research.

Koekebakker and Sodal (2002) developed an equilibrium-based real options model of an operating electricity production unit whose supply is given by a stochastic mean-reverting process. Hlouskova et al. (2002) implemented a real options model for the unit commitment problem of a single turbine in a liberalized market. Price uncertainty was captured by a mean-reverting process with jumps and time-varying means to account for seasonality. Rocha, Moreira, and David (2002) studied the competitiveness of thermopower generation in Brazil under current regulations and used real options to assess how to motivate private investment in thermopower.

Wang (2003) applied NPV, NPV with simulation, and binomial options pricing model to study a case on Yalongjiang River basin development. More specifically, a deferral option of Project 1 is studied. When doing the real options analysis, the thesis compared the usage of NPV and electricity price as the underlying, and found that the electricity price is a more appropriate underlying for the options analysis. The thesis concluded that real options analysis using the electricity price as underlying is an appropriate method for valuing the deferral options of Project 1 and similar hydropower projects.

Other real options applications

Nichols (1994) interviewed Merck CFO Judy Lewent to unravel the secrets of Merck's successful risk management of its R&D portfolios, and the crux of the success is real options. Faulkner (1996) warns R&D managers that traditional DCF valuation could miss important sources of value of R&D projects, and real options valuation can fix such a bias. Childs and Triantis (1999) examined dynamic R&D policies and the valuation of R&D projects in a real options framework. Neely and de Neufville (2001) proposed using hybrid real options valuation method that combines the best features of decision analysis and real options to value risky product development projects.

2.4.3. Real Options "in" and "on" projects

Real options can be categorized as those that are either "on" or "in" projects (de Neufville, 2002). Real options "on" projects are financial options taken on technical things, treating technology itself as a 'black box'". Real options "in" projects are options created by changing the actual design of the technical system. For example, de Weck et al (2004) evaluated real options "in" satellite communication systems and determined that their use could increase the value of satellite communications systems by 25% or more. These options involve additional fuel on satellites in order to achieve a flexible design that can adjust capacity according to need.

One dimension of the general development of options is depicted in Figure 1-2. With the development of options theory, the scope of application is expanding, from financial options to real options "on" projects to real options "in" projects. Real options "in" projects further expand the options thinking into physical systems, adding flexibility systematically with awareness. With the success of options theory and its key insights into uncertainty, it has bright prospects to improve engineering systems design in meeting customer demands, economical feasibility or profitability, and regulatory requirements.

In general, real options "in" systems require a deep understanding of technology. Because such knowledge is not readily available among options analysts, there have so far been few analyses of real options "in" projects, despite the important opportunities available in this field. Moreover, because the data available for real options "in" project analysis is of much poorer quality than that of financial options or real options "on" projects, real options "in" projects are different and need an appropriate analysis framework - existing options theory has to adapt to the new needs of real options "in" projects.

The real options discussed in previous sections were all real options "on" projects. There is much less literature on real options "in" projects. The following is a discussion of literature on real options "in" projects from de Neufville, Sholtes, and Wang (2005): "Zhao and Tseng (2003) discussed the value of flexibility in infrastructure facilities. Enhancing the foundation requires extra up-front cost, but has a return for future expansion when uncertain elements are realized to be good. This trade-off can be viewed as an option in which a premium has to be paid first and the option can be exercised later. They used an example of construction of a public garage to illustrate the class of problem. Trinomial lattice and stochastic dynamic programming were used to model the demand and optimal expansion process. A model with flexibility is compared with that without flexibility, and the difference of the optimal value from the two models is the value of flexibility. This value of flexibility is significant in the case. Zhao, Sundararajan, and Tseng (2004) presented a multistage stochastic model for decision making in highway development that incorporated real options in both the development and operations phase. A simulation algorithm based on the Monte Carlo simulation and least- squares regression is developed. Ho and Liu (2003) presented a quantitative valuation method based on options pricing theory for evaluating major investments in emerging architecture/engineering/construction (A/E/C) technology investments. The framework took into account technology investment risks and managerial options. Leviakangas and Lahesmaa (2002) discussed the application of real options in evaluation of intelligent transportation system and pointed out the shortcoming of traditional cost-benefit analysis that may discard the value of real options. Kumar (1995) presented the real options approach to value expansion flexibility and illustrated its use through an example on flexible manufacturing systems. Ford, Lander, and Voyer (2002) proposed a real options approach for proactively using strategic flexibility to recognize and capture project values hidden in dynamic uncertainties. An example for a toll road project is employed in their work.

The existing literature on real options "in" projects does not provide a generic framework for real options "in" projects, but on single specific projects or issues. They do not attack the general issues facing real options "in" projects, for example, the path-dependency issue or the identification of real options. This area needs a lot of creative work.

The author searched the literature extensively, and mentioned all real options "in" projects papers he found in this section. Compared to the abundance of sections 2.4.1. and 2.4.2. that treat real options "on" projects, the conclusion is that existing work on real options "in" projects is very limited.

2.5. Conclusions from the literature search

Following the three threads in Figure 1-1, extensive literature search reveals conclusions regarding each of the three threads:

- It is an important yet underdeveloped subject to design engineering systems with proactive management of uncertainties;
- Real options "in" projects is an important new area to explore where limited work exists;

- And, mathematical programming, especially mixed integer programming, has interesting prospects to deal with water resources planning problems and stochastic problems (options analysis is a stochastic problem).

Chapter 3 Standard Options Theory

The modern options theory was founded by Black, Scholes, and Merton (1973). Gradually, options methodology and thinking has been gradually extended to broader areas in finance and non-finance. Its insights into uncertainty and flexibility enhance the ability of human beings to deal with forever-changing environments.

3.1. Financial Options

There are two basic types of options: calls and puts. A call option gives the holder the right to buy an underlying asset for a specified exercise price within or at a specified time. A put option gives the holder the right to sell the underlying under similar circumstances. Expiration date is also called maturity. Exercise price is also called strike price.

Financial options are also categorized by the time when they can be exercised. American options can be exercised at any time up to the expiration date. European options can be exercised only on the expiration date.

The underlying assets for financial options include stocks, stock indices, foreign currencies, debt instruments, commodities, and futures contracts. Besides the real options, this thesis is only discussing the financial options built on underlying assets of stocks, or stock options.

Example of a stock call option:

John buys one European stock call option contract on Lucent stock with a strike price of $1.50. Suppose the current price of Lucent is $1.30, the expiration date is in three months. Because the option is European, John can exercise the option only on the expiration date. If the stock price on the expiration date is less than $1.50, John will choose not to exercise. If the stock price on the expiration date is greater than $1.50, John will choose to exercise. For instance, if the stock price on the expiration date is $1.45, John will not exercise the option, he can buy a share of stock directly on the market for $1.45, $0.05 less than the exercise price of $1.50. If the stock price on the expiration date is $1.60, John will exercise the option and earn $0.10 because he can immediately sell the stock that he buys for $1.50.

Key Property of an Option

The holder of an option has the right to exercise the option, but no obligation to exercise the option. The key property of an option is the asymmetry of the payoff, an option holder can avoid downside risks and limit the loss to the price of getting the option, while she can take advantage of the upside risks and the possible gain is unlimited. See Figure 3-1 for the example of a stock option.

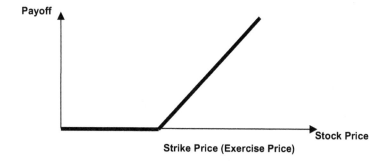

Figure 3-1 Payoff Diagram for a Stock Call Option

For the above stock call option in Figure 3-1, if the current stock price is lower than the strike price, people would not exercise it, the loss is limited to the price to get the option; if the current stock price is higher than the strike price, people would exercise it, and the payoff is the current stock price minus the strike price. There is no upper bound of the payoff but the lower bound of the payoff is zero, so asymmetry exhibits. The maximum loss is equivalent to the original purchase price of the option.

3.2. Cornerstones for Options Valuation

The value of an option is not straightforward, and it is an interesting question how to value an option objectively. The cornerstones for the modern stock options valuation models are two assumptions: no arbitrage and Brownian motion of stock price.

3.2.1. No Arbitrage

Arbitrage involves getting profit by simultaneously entering into transactions in 2 or more markets. See the following example: Considering a stock that is traded on both the New York Stock Exchange and the London Stock Exchange. If the stock price is \$17.7 in New York and £10 in London when the exchange rate is \$1.8000 per pound. An arbitrager could simultaneously buy 1000 shares of the stock in New York and sell them in London to obtain a risk-free profit of

$$1000 \times [(\$1.8 \times 10) - \$17.7]$$

or $300. Arbitrage opportunities such as the one just described cannot last for long. As arbitrageurs buy the stock in New York, the forces of supply and demand will cause the stock price to rise. Similarly, as arbitrageurs sell the stock in London, the forces of supply and demand will cause the stock price to drop. Very quickly, the two prices will be equivalent at the current exchange rate. Indeed, the existence of profit hungry arbitrageurs makes it unlikely that a major price disparity could ever exist in the first place.

If no arbitrage opportunity exists, a portfolio of the stock and the stock option can be set up in such a way that there is no uncertainty about the value of the portfolio. Because the portfolio has no risk, the return earned on it must equal the risk-free interest rate.

A riskless portfolio consisting of a position in the option and a position in the underlying stock is created. In the absence of arbitrage opportunities, the return from the portfolio must be the risk-free interest rate. The reason why a riskless portfolio can be created is the stock price and the option price are both affected by the same courses of uncertainty: stock price changes. In a short period of time, when an appropriate portfolio is established, the gain or loss from the stock option is always offset by the loss or gain from the stock position so that the value of the portfolio is known with certainty at the end of the short period of time to earn a risk-free rate of interest. For that short period of time, the price of a call option is perfectly positively correlated with the price of the underlying stock, and the price of a put option is perfectly negatively correlated with the underlying stock.

For a simple example: A stock price is currently $10, and it is known that at the end of a period of y months the stock price will be either $11 or $9. There is a European call option to buy the stock for $10.5 at the end of y months. This option will have one of two values at the end of the y months. If the stock price turns out to be $11, the value of the option will be $0.5; if the stock option turns out to be $9, the value of the option will be 0. The situation is illustrated in Figure 3-2:

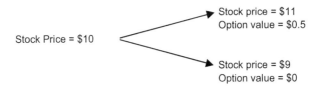

Figure 3-2 Stock Price Movement in Numerical Example

Consider a portfolio consisting of a long position[1] in x shares of the stock and a short position[2] in one call option. How to calculate the value of x that makes the portfolio riskless? If the stock price moves up from $10 to $11, the value of the share is 11x and the value of the call option for purchasing a single share at $10.5 is $0.5, so that the total value of the portfolio is 11x – 0.5. If the stock price moves down from $10 to $9, the value of the shares is 9x and the value of the option is 0, so that the total value of the portfolio is 9x. The portfolio is riskless if the value of x is chosen so that the final value of the portfolio is the same for both cases. This means:

$$11x - 0.5 = 9x$$

or

$$x = 0.25$$

If the stock price moves up to $11, the value of the portfolio is

$$11 \times 0.25 - 0.5 = 2.25$$

If the stock price moves down to $9, the value of the portfolio is

---

[1] A long position is to buy the underlying asset on a certain specified future date for a certain specified price.
[2] A short position is to sell the underlying asset on a certain specified future date for a certain specified price.

$$9 \times 0.25 = 2.25$$

Regardless of whether the stock price moves up or down, the value of the portfolio is always $2.25 after y months.

A riskless portfolio must, in the absence of arbitrage opportunities, earn the risk-free rate of interest of r. It follows that the value of the portfolio today must be equivalent to the continuously compounded present value of 2.25, or

$$2.25e^{-\frac{ry}{12}}$$

The value of the stock price today is known to be $10. Suppose the option price is f. The value of the portfolio is

$$10 \times 0.25 - f = 2.25e^{-\frac{ry}{12}}$$

or

$$f = 2.5 - 2.25e^{-\frac{ry}{12}}$$

3.2.2. Brownian motion and Wiener Processes

The standard model for stock prices is a geometric Brownian motion with constant volatility. Standard Brownian motion is one of the most important basic notions of stochastic processes, and in particular, is the basis of modern options theory.

To develop a sound theory of option pricing, one should describe the stock price evolution using a dynamic model with a reasonable agreement with reality. The exact formulation of the model for stock price evolution was a subject of debates for over a century.

Brownian motion originally refers to the random motion observed under microscope of a pollen immersed in water. Albert Einstein pointed out that this motion is caused by random bombardment of heat-excited water molecules on the pollen. More precisely, each of the pollen's steps (in both x- and y- directions) is an independent normal random variable.

Albert Einstein developed the notions of Brownian motion in the beginning of the 20th century. In 1905 he defended his Ph.D. thesis on the subject of the separation of two large particles experiencing random hits from surrounding small molecules. For this work he received the Nobel Prize (Ironically, he did not receive the Nobel Price for the Theory of Relativity). Although he himself considered his work not particularly important, this work laid the ground for the theoretical understandings and beginnings of stochastic processes altogether. Further contributors to the subject were Markov, Uhlenbeck, Khintchine, Wiener, Smoluchowski, Ito, and Stratonovich. It was only in the 1960s that the theory of Brownian motion was applied to modeling stock prices.

Stock prices are influenced by an astronomical number of independent random factors together. Each factor is trivial in the total influence. This kind of random variables, stock prices in this case, usually follows normal distribution approximately. The reason why a normal distribution is not used to describe stock prices is because stock prices cannot be negative. A lognormal distribution describes the rate of change of stock prices (expressed using continuous compounding) to be normal. The change of a stock price can be negative, which means the effective market price is decreasing while it is still positive.

Consider the following discrete construction. Let Z_{t_0} be the position of a particle at time t_0. Let at time $t_0 + \Delta t$ the position of the particle be $Z_{t_0} + \Delta Z$, where the increments are related:

$$\Delta Z = \varepsilon \sqrt{\Delta t}$$

where ε denotes a random sample from a standard normal distribution (mean 0 and standard deviation 1).

$$Z_{t_n} = Z_{t_{n-1}} + \varepsilon_n \cdot \sqrt{\Delta t}$$

$$t_n - t_{n-1} = \Delta t$$

Compounding n such increments, one can get for a finitely large interval of time $T = n\Delta t$:

$$Z(t_0 + T) = Z_{t0} + \sum_{i=1}^{n} \varepsilon_i \sqrt{\Delta t}$$

Here ε_i are all independent samplings from a standard normal distribution. Considering the limit of $\Delta t \to 0_+$, it may be shown that the resulting process converges to a limit, which is called standard Brownian motion, and is also referred to as a Wiener process.

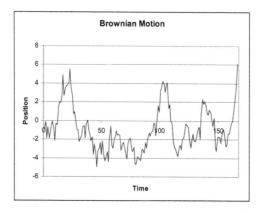

Figure 3-3 One Path of Brownian Motion (Δt = 1)

Figure 3-3 exhibits a single path of a standard Brownian motion with initial condition $Z_0 = 0$.

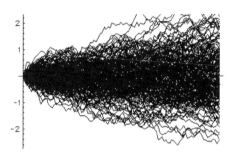

Figure 3-4 Two Hundred Paths of Brownian Motion

Figure 3-4, in turn, shows two hundred paths of the standard Brownian motion.

The basic Wiener process, dz, has a drift rate (i.e. average change per unit of time) of zero and a variance of 1.0. The drift rate of zero means that the expected value of z at any future time is equal to its current value. The variance rate of 1.0 means that the variance of the change in z in a time interval of length T equals T.

A generalized Wiener process for a variable x can be defined in terms of dz as follows:

$$dx = adt + bdz$$

Equation 3-1

where a and b are constants, dz is the basic Wiener process. The adt term implies that x has an expected drift rate of a per unit of time. This means that, without the bdz term, in a period of time of length T, x increases by an amount of aT. The bdz term can be regarded as adding noise or variability to the path followed by x. The amount of this noise or variability is b times a basic Wiener process. A basic Wiener process has a standard deviation of 1.0. It follows that *b* times a Wiener process has a standard deviation of *b*.

In a small time interval Δt, the change in the value of x, Δx, is

$$\Delta x = a\Delta t + b\varepsilon\sqrt{\Delta t}$$

Since ε is a random number drawing from a standard normal distribution. Thus, Δx has a normal distribution with

$$\text{Mean of } \Delta x = a\Delta t$$
$$\text{Standard deviation of } \Delta x = b\sqrt{\Delta t}$$
$$\text{Variance of } \Delta x = b^2\Delta t$$

Stock Price Process Model

It is usually assumed that asset prices follow Geometric Brownian Motion where the logarithm of the underlying variable follows a generalized Wiener process.

If the price of a non dividend paying stock, S, follows geometric Brownian motion:

$$dS = \mu Sdt + \sigma Sdz \qquad\qquad \text{Equation 3-2}$$

where S is the stock price, μ is the expected return on the stock, and σ is the volatility of the stock price. The volatility of a stock price can be defined as the standard deviation of the return provided by the stock in one year when the return is expressed using continuous compounding. The volatility is also the standard deviation of the natural logarithm of the stock price at the end of one year. Both definitions are equivalent.

Then, using Ito's lemma (see Appendix 3A) to get:

$$d \ln S = (\mu - \frac{\sigma^2}{2})dt + \sigma dz \qquad \text{Equation 3-3}$$

From this equation, the variable $\ln S$ follows a generalized Wiener process, the change in $\ln S$ between 0 and t is normally distributed, so that S has a lognormal distribution.

Lognormal Properties of stock price

In general, a lognormal distribution probability density function is as follows:

$$f(x) = \frac{1}{x} \cdot \frac{1}{\sigma\sqrt{2\pi t}} \cdot e^{\frac{-(\ln x - \mu)^2}{2\sigma^2 t}} \qquad \text{Equation 3-4}$$

but the following will show a more intuitive explanation of the lognormal distribution of the stock prices. See Figure 3-5 for the shape of lognormal density function.

The lognormal distribution of price means the logarithm of the price has a normal distribution. To illustrate, if a stock is priced at $100 per share and prices have a normal distribution, the distribution of prices is the familiar bell-shaped curve centered at $100, but if the prices have a lognormal distribution, then it is the logarithm of the price which has a bell-shaped distribution about $\ln(100) = 4.605$. The logarithm of the prices is equally likely to be 5.298 or 3.912, i.e., 4.605 ± 0.693 corresponding to prices of $200 and $50, respectively. If the lognormal probability density curve is plotted as a function of price rather than as a function of the logarithm of price, the curve will appear positively skewed with tails more nearly depicting the observed behavior of stock prices.

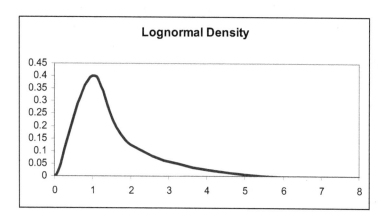

Figure 3-5 Lognormal Density

Lognormality arises from the process of return compounding, in other words, the lognormal property of stock prices applies when the return rate earned on a stock between time 0 and t is continuously compounded. It is important to distinguish the continuously compounded rate of return and the annualized return with no compounding as $\frac{1}{t}(\frac{S_t - S_0}{S_0})$.

3.3. Options Valuation Tools

The first model to calculate options value is the Black-Scholes formula, which is sometimes regarded as arcane. Interests in option pricing, however, has picked up in recent years as more powerful computers can aid very sophisticated model building.

With simulation methods available easily as Excel add-ins or more professional alternatives such as @Risk or Crystal Ball, people are able to do a hundred thousand simulations easily and get the payoff distribution as well as the value of real options[1]. Besides, the binomial model proves very successful in option pricing and decision analysis is sometimes another approach to value options approximately[2].

Five inputs are needed for an options valuation (if considering the simplest situation when there is no dividend):

-   strike price
-   risk-free interest rate
-   time to expiration
-   current stock price
-   uncertainty (with volatility as the measurement)

Among the five inputs of an options model, the first four inputs are relatively easier to get, while the last one uncertainty, which is estimated by volatility in most cases, is more difficult to estimate. With a lot of historical data available on the stock market, it is relatively trivial to get σ for a stock option. However, for a real option, lack of historical data is a common problem except a few specific industries, such as pharmaceutical industry. Because of the lack of historical data, it is very hard to justify the choice of volatility. This is one of the practical difficulties facing real options valuation methods.

3.3.1. The Black-Scholes Model

The Black-Scholes-Merton analysis is based on the no-arbitrage condition.

[1] Simulation generates values of uncertain variables according to the probability distribution of the variables, uses those values as inputs, and predicts the output. With a great number of repetitions, the probability distribution of the output is established.
[2] The valuation by decision analysis is not real options valuation strictly because it does not use the risk-neutral valuation (refer to 3.4.).

The stock price process is the one as we developed in the last section as Equation 3-2:

$$dS = \mu S dt + \sigma S dz$$

Suppose $f(S,t)$ is the price of a call option, which is some function of the stock price of S and time of t. Hence from Ito's Lemma (see Appendix 3A):

$$df = (\frac{\partial f}{\partial S} \mu S + \frac{\partial f}{\partial t} + \frac{1}{2} \frac{\partial^2 f}{\partial S^2} \sigma^2 S^2) dt + \frac{\partial f}{\partial S} \sigma S dz \qquad \text{Equation 3-5}$$

The Wiener processes dz underlying f and S are the same. It follows that by choosing a portfolio of the stock and the stock option, the Wiener process can be eliminated. The appropriate portfolio is short one call option and long an amount $\partial f / \partial S$ of shares. Define $\prod$ as the value of the portfolio. By definition

$$\prod = -f + \frac{\partial f}{\partial S} S \qquad \text{Equation 3-6}$$

Note the portfolio is riskless only for an infinitesimally short period of time. As S and t change, $\partial f / \partial S$ also changes. To keep the portfolio riskless, it is necessary to constantly change the composition of the portfolio.

Because the discrete version of equations Equation 3-5 and Equation 3-2 are

$$\Delta f = (\frac{\partial f}{\partial S} \mu S + \frac{\partial f}{\partial t} + \frac{1}{2} \frac{\partial^2 f}{\partial S^2} \sigma^2 S^2) \Delta t + \frac{\partial f}{\partial S} \sigma S \Delta z$$

and

$$\Delta S = \mu S \Delta t + \sigma S \Delta z$$

Then the change $\Delta \Pi$ in the time interval Δt is given by

$$\Delta \Pi = -\Delta f + \frac{\partial f}{\partial S} \Delta S$$
$$= -(\frac{\partial f}{\partial S} \mu S + \frac{\partial f}{\partial t} + \frac{1}{2} \frac{\partial^2 f}{\partial S^2} \sigma^2 S^2) \Delta t - \frac{\partial f}{\partial S} \sigma S \Delta z + \frac{\partial f}{\partial S} \mu S \Delta t + \frac{\partial f}{\partial S} \sigma S \Delta z \qquad \text{Equation 3-7}$$
$$= (-\frac{\partial f}{\partial S} - \frac{1}{2} \frac{\partial^2 f}{\partial S^2} \sigma^2 S^2) \Delta t$$

The equation does not involve Δz, the portfolio must be riskless during time Δt under the assumption of no arbitrage. It follows that

$$\Delta \Pi = r \Pi \Delta t \qquad \text{Equation 3-8}$$

where r is the risk-free interest rate. Substituting from equations Equation 3-6 and Equation 3-7, this becomes

$$(\frac{\partial f}{\partial t} + \frac{1}{2} \frac{\partial^2 f}{\partial S^2} \sigma^2 S^2) \Delta t = r(f - \frac{\partial f}{\partial S} S) \Delta t$$

So that

$$\frac{\partial f}{\partial t} + rS \frac{\partial f}{\partial S} + \frac{1}{2} \sigma^2 S^2 \frac{\partial^2 f}{\partial S^2} = rf \qquad \text{Equation 3-9}$$

Equation 3-9 is the Black-Scholes-Merton differential equation. Its solution depends on the boundary conditions used. In case of a European call option, the key boundary condition is

$$f = \max[S - X, 0] \quad \text{when } t = T$$

Solving the differential equation subject to the boundary conditions to get the Black-Scholes formulas for the prices at time zero of a European call option on a non-dividend-paying stock:

$$c = S_0 N(d_1) - X e^{-rT} N(d_2)$$

Equation 3-10

where

$$d_1 = \frac{\ln(S_0 / X) + (r + \sigma^2 / 2)T}{\sigma \sqrt{T}}$$

$$d_2 = \frac{\ln(S_0 / X) + (r - \sigma^2 / 2)T}{\sigma \sqrt{T}} = d_1 - \sigma \sqrt{T}$$

and N(x) is the cumulative probability distribution function for a variable that that is normally distributed with a mean of zero and a standard deviation of 1.0.

In case of a European put option, the key boundary condition is

$$f = \max[X - S, 0] \quad \text{when } t = T$$

solving the differential equation to get the Black-Scholes formula for the prices at time zero of a European put option on a non-dividend-paying stock

$$p = X e^{-rT} N(-d_2) - S_0 N(-d_1)$$

After going through all the derivation of the Black-Scholes formula, the two most important assumptions under the options pricing are stressed again here:

- no arbitrage, and
- geometric Brownian motion.

It is key to understand the role of these two assumptions in real options valuation. And thus understand the applicability for real options in different scenarios.

3.3.2. Valuation by Simulation

Simulation models compute typically thousands of possible paths of the evolution of the value of the underlying from today to the final day studied. With options decision rule imbedded in each of the paths, the expected value on the final day is discounted back to today to obtain the options value.

For example, the current price of a stock is $20, volatility is 30% per year, risk-free rate is 5% per year. A European stock call option is built on such a stock with time to expiration of 3 months and strike price of $22.

Simulating the stock price 3 months later to get the distribution of the stock price yields Figure 3-6. With the European call option imbedded, the distribution of prices less than $22 is cut because people won't exercise it in such a case. See Figure 3-7. The expected of the chunk of the distribution greater than $22 minus $22 is the future value of the option. The future value of the option is then discounted to get the option value.

One of the advantages of the simulation model is that it can handle path-dependent options, in which the value of options depends not only on the value of the underlying, but also on the particular path followed by the underlying.

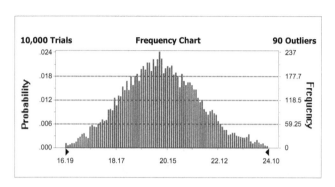

Figure 3-6 Distribution of Stock Price for Valuation by Simulation

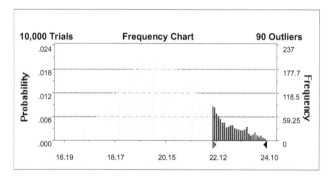

Figure 3-7 Distribution of Exercise of Option for Valuation by Simulation

With the fast development of computer hardware and software technologies, simulation models have been more and more powerful and easy-to-use. A normal laptop can run thousands of simulations in seconds, and software programs such as Crystal Ball make the simulation method accessible to everybody.

An example of options valuation by simulation is provided in section 3.3.4.

3.3.3. Binomial Real Options Valuation

Considering a stock whose price is initially S_0 and an option on the stock whose current price is f. Suppose that the option lasts for time T. During the life of the option, the stock price can either move up from S_0 to a new level, S_0u, or down from S_0 to a new level S_0d. The proportional increase in the stock price when there is an up movement is $u - 1$; the proportional decrease when there is a down movement is $1 - d$. If the stock price moves up to S_0u, the payoff from the option is assumed to be f_u; if the stock price moves down to S_0d, the payoff from the option is assumed to be f_d. Figure 3-8 illustrates the situation:

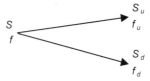

Figure 3-8 Stock and Option Price in a One-step Binomial Tree

Considering a portfolio consisting of a long position of x shares and a short position in one option. If there is an up movement in the stock price, the value of the portfolio at the end of the life of the option is

$$S_0ux - f_u$$

If there is a down movement in the stock price, the value becomes

$$S_0dx - f_d$$

The two are equal when

$$S_0 ux - f_u = S_0 dx - f_d$$

or

$$x = \frac{f_u - f_d}{S_0 u - S_0 d}$$ Equation 3-11

In this case, the portfolio is riskless. Due to the no arbitrage condition, the portfolio must earn risk-free interest rate. x is the ratio of the change in the option price to the change in the stock price.

The present value of the portfolio is

$$(S_0 ux - f_u)e^{-rT}$$

where r is the risk-free interest rate. The cost of establishing the portfolio is

$$S_0 x - f$$

It follows that

$$S_0 x - f = (S_0 ux - f_u)e^{-rT}$$

or

$$f = S_0 x - (S_0 ux - f_u)e^{-rT}$$

Substituting x from Equation 3-11, this equation reduces to

$$f = e^{-rT}[p_u f_u + p_d f_d]$$

Equation 3-12

where

$$p_u = \frac{e^{rT} - d}{u - d}$$

Equation 3-13

$$p_d = 1 - p_u$$

Equation 3-14

One way to match volatility with u and d is

$$u = e^{\sigma\sqrt{\Delta t}}$$

Equation 3-15

$$d = e^{-\sigma\sqrt{\Delta t}}$$

Equation 3-16

One more step can be added to the binomial tree as Figure 3-9.

Repeated application of   Equation 3-12 gives

$$f_u = e^{-rT}[pf_{uu} + (1-p)f_{ud}]$$
$$f_d = e^{-rT}[pf_{ud} + (1-p)f_{dd}]$$

And finally get:

$$f = e^{-rT}[pf_u + (1-p)f_d]$$

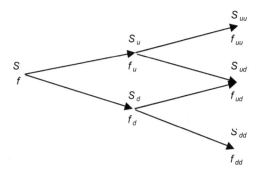

Figure 3-9 A Two-Step Binomial Tree

3.3.4. Options Valuation and Decision Tree Analysis

Options analysis is continuous, but decision trees are discrete. Normally, the decision tree analysis will not give the correct value for options because it is not a risk neutral analysis. Decision tree analysis does not refer to arbitrage-enforced price, and uses the actual probabilities of the price movement of the underlying assets. If using risk-neutral evaluation and simulation, however, the decision tree analysis will give the exactly same answer as Black-Scholes.

To a certain extent, what the options theory offers is an understanding of the stock prices that are lognormally distributed. From today's stock price and the volatility of the stock price, the distribution of a stock price at a future date can be derived. With the understanding of the stock price distribution, decision tree analysis can be applied.

An Example of Option Valuation

Assume that there is a stock, where stock price S_0 is $20 now, the volatility σ is 30%, risk-free interest rate r is 5%. A call option is built on this stock, the strike price X is $22, and the time to maturity T is 1 year.

Applying two methods to get the value of this call option.

Method 1: Black-Scholes Formula

The Black-Scholes Formula is as Equation 3-10

$$c = S_0 N(d_1) - Xe^{-rT} N(d_2)$$

where

$$d_1 = \frac{\ln(S_0/X) + (r + \sigma^2/2)T}{\sigma\sqrt{T}}$$

$$d_2 = \frac{\ln(S_0/X) + (r - \sigma^2/2)T}{\sigma\sqrt{T}} = d_1 - \sigma\sqrt{T}$$

and $N(x)$ is the cumulative probability distribution function for a variable that that is normally distributed with a mean of zero and a standard deviation of 1.0.

Substitute the actual value of S_0, σ, r, X, and T into the formula, and the value of this call option is $1.994.

Method 2: Decision Tree Analysis using Monte Carlo Simulation

The basic structure of the tree is as Figure 3-10:

Figure 3-10 Decision Tree for Options Valuation

The rectangle is a decision point at the expiration day, at which there are two possible decisions: exercise the option, and the value of the option is the stock price then minus the strike price of $22; do not exercise the option, the value of the option is 0 in this case.

Now the key of the above decision tree analysis is the stock price. The stock price is a stochastic process. It has a specific distribution on the expiration day that can be derived from the current stock price, volatility, risk-free rate, and time to expiration. The assumptions needed for that deduction are the cornerstones of modern finance theory, i.e. the geometric Brownian motion of the stock price and the no-arbitrage assumption. These two assumptions also lead to the Black-Scholes formula.

The two assumptions lead to the stock price on the expiration day following a lognormal distribution with an expected value as Equation 3-18

$$E(S_T) = S_0 e^{rT}$$

where T is time to expiration and r is the risk-free rate.

Substitute the actual value of S_0, r, and T into Equation 3-18 to get the expected value of the stock price on the expiration day after a year:

$$E(S_T) = \$21.025$$

Two parameters, i.e. μ and σ, are needed to specify the lognormal distribution of the stock price on the expiration day. Please refer to Equation 3-4. In this example, σ is 0.3. μ is the expected value of the annual return expressed using continuous compounding, and

$$\mu \neq \ln[E(S)],$$

but

$$\mu = E[\ln(S / S_0)] = \ln[E(S)] - \sigma^2 / 2 \qquad \text{Equation 3-17}$$

or

$$\mu = \ln(21) - (0.3)^2 / 2 = 3.001.$$

With the value of μ and σ, the distribution of the stock price on the expiration day is specified. The last important thing is that the option is to expire a year later, but the value of the option as of today needs to be calculated. So the expected value obtained by the decision tree needs to be discounted.

Finally, Monte Carlo simulation is applied to get the value of the option c. The software used is Crystal Ball. The relative precision of the simulation of the mean is set to be 1%. It means that the software will stop simulation after the mean of the simulated results is within ±1% range of true expected mean.

The software simulates 249,500 times before it stops and reaches the relative precision of 1%. Please see the output from Crystal Ball as Table 3-1:

The result of Table 3-1 shows the true expected value or the value of the call option should be in the range $2.00082 ± 1% x $2.00082, or ($1.981, $2.021). The result from the Black-Scholes formula is $1.994, exactly in the range. This test shows that the expected value of the tree (after discounting) is the call option value and it is the same as the result from Black-Scholes formula. The precision of the simulation can be improved, even though the mean of the simulated results is fluctuating around the expected value.

Table 3-1 Option Valuation by Decision Tree Results

Statistic	Value	Precision
Trials	249,500	
Mean	**2.00082**	1.00%
Median	0.00000	
Mode	0.00000	
Standard Deviation	3.87972	0.88%
Variance	15.05222	
Skewness	2.72981	
Kurtosis	12.53695	
Coeff. of Variability	1.93906	
Range Minimum	0.00000	
Range Maximum	50.71999	
Range Width	50.71999	
Mean Std. Error	0.00777	

This example shows an interesting result that options can be valued by decision tree analysis with simulation. In some sense, modern finance theory helps people to get the distribution of the stock price at any future day with the observable parameters of the current stock price, the risk-free rate, and the volatility of the stock price.

3.4. Risk-neutral Valuation

Note in the Black-Scholes formula (Equation 3-10), the value of an option does not depend on discount rate or other variables that are affected by the risk preferences of investors. The variables presented in the formula – current stock price, time, stock price volatility, and the risk-free rate of interest – are all independent of risk preferences. This leads to the most important tool for the analysis of options and other derivatives, risk neutral valuation.

By risk-neutral valuation, we assume all investors are risk-neutral and do not need compensation for taking risks. In this risk neutral world, the expected return from the underlying asset is equal to the risk-free interest rate, and the discount rate to discount the expected payoff is risk-free interest rate. However, the solutions obtained in a risk-neutral world are valid in all world, not just where investors do not request risk-premium,

To illustrate risk-neutral valuation further, we refer to a simple binomial model structure as Figure 3-8, the expect stock price at time T, $E(S_T)$, is given by

$$E(S_T) = p_u S_0 u + (1 - p_u) S_0 d$$

or

$$E(S_T) = p_u S_0 (u - d) + S_0 d$$

Substituting from Equation 3-13, this reduces to

$$E(S_T) = S_0 e^{rT}$$ Equation 3-18

The expected growth of the stock price is the risk-free rate. Setting the probability of the up movement equal to p_u is to assume that the expected return on the stock is the risk-free rate.

In a risk-neutral world, all individuals are expected value maximizers and require no compensation for risk, and the expected return on all securities is the risk-free rate. Equation 3-12 shows that the value of an option is its expected payoff in a risk-neutral world discounted at the risk-free rate. The risk-neutral valuation principle states that it is valid to assume the world is risk neutral when pricing options. The result is correct for all worlds, not only in the risk-neutral world.

3.5. Exotic options

Exotic options are derivatives with more complicated payoffs than the standard European or American options. Exotic options valuation techniques are of special interests to real options researchers because real options usually possess more a complicated payoff structure than standard European or American options.

In this section we will first introduce a number of exotic options, and then describe briefly the current valuation techniques for several exotic options that are relevant to this book.

3.5.1. Types of exotic options

The categorization of exotic options presented in this book follows Hull [1999].

Packages

A package is a portfolio of standard European options, forward contracts, cash, and the underlying asset itself. There are different types of packages, such as bull spreads, bear spreads, butterfly spreads, calendar spreads, straddles, and strangles. For details of the packages, refer to Hull (1999).

Nonstandard American Options

For a standard American option, exercise can take place at any time during the life of the option and the exercise price is the same. For a nonstandard option, the exercise may be restricted to a certain period of time or the strike price can be changing. For example, Bermudan option restricts its early exercise to certain dates during the life of the option. A bond option that can be exercised only on coupon payment dates is a Bermudan option. Another example, a warranty issued by a company on its own stock sometimes can only be exercised during only part of its life, and sometimes the strike price increases as time passes.

Forward Start Options

Forward start options are options that will start at some time in the future. They are sometimes used in incentive options. The terms of the options will usually specify when it starts it will be at the money.

Compound options

Compound options are options on options. There are four main types of compound options: a call on a call, a put on a call, a call on a put, and a put on a put. Considering a call on a call, on the first the exercise date, the holder has the right not the obligation to buy the second call option at a first exercise price, and if the holder buys the second call option, the holder has the right not obligation to buy the underlying asset at the second exercise date for a second exercise price. Many real options have the compound options features.

Chooser options

A chooser option allows the holder to choose whether the option is a call or a put after a predetermined period of time. The value of the chooser option at the time of choice is

$$max(call, put)$$

Barrier options

Barrier options are options where the payoff depends on whether the underlying asset's price reaches a certain level during a certain period of time. Barrier options can be categorized as knock-out options or knock-in options. A knock-out option stops to exist when the underlying asset price reaches a certain barrier; a knock-in option begins to exist when the underlying asset price reaches a certain barrier. For example, a down-and-out call is a regular call option that stops to exist when the underlying asset price reaches a certain lower bound. Another example, a down-and-in call option is a regular option that comes into existence only if the stock price reaches a certain lower bound. Barrier options are path-dependent, and how to value path-dependent options is especially interesting to real options researchers.

Binary options

Binary options have discontinuous payoffs. For example, a cash-or-nothing call pays off nothing if the stock price ends up below the strike price at the expiration, and pays a fixed amount if the stock price ends up above the strike price. A asset-or-nothing call pays off nothing if the stock price ends up below the strike price at the expiration, and pays off an amount equal to the stock price if it ends up above the strike price. A European option can be thought of as equivalent to a long position in an asset-or-nothing call and a short position in a cash-or-nothing call, where the cash payoff of the cash-or-nothing call is equal to the strike price.

Lookback options

Lookback options are another kind of path-dependent options. The payoffs of lookback options are the maximum or minimum stock price reached during the life the option. A lookback call can help the holder buy the underlying asset at the lowest price achieved during the life of the option; and a lookback put can help the holder sell the underlying asset at the highest price achieved during the life of the option.

Shout options

A shout option holder can "shout" at one time during the life of the option, and the holder gets the maximum of the usual payoff from an ordinary European option or the value at the time of the shout. For example, the strike price is $30 and the holder of a call when the price of the underlying asset is $40. If the final asset price is less than $40 then the holder receives $10; if the final asset price is greater than $40 then the holder receives the excess of the asset price over $30.

Asian options

Asian options are options where the payoff depends on the average price of the underlying asset during part or all life of the option. The payoff from an average price call is $\max(0, S_{average} - X)$, that from an average price put is $\max(0, X - S_{average})$, that from an average strike call is $\max(0, S_T - S_{average})$, and that from an average strike put is $\max(0, S_{average} - S_T)$, where $S_{average}$ is the average value of the underlying asset calculated over a period of time, X is the strike price, and S_T is the stock price at maturity. Asian options are less expensive than regular options and are arguable more appropriate to meet the needs of corporate treasurers. Suppose a multinational corporation headquartered in Boston expects to receive a cash flow in Euros spread in the next year, the treasurer is probably more interested in guaranteeing the average exchange rate realized during the next year is above some minimum level. Asian options are among the most difficult path-dependent financial options in terms of valuation.

Options involving several assets

Options involving two or more risky assets are sometimes called rainbow options as well. For example, the bond futures contract traded on the Chicago Board of Trade allows the seller to choose between a large number of different bonds when delivering.

Basket options

A basket option depends on the underlying of a portfolio of assets. The assets are usually stocks, stock indices, or currencies.

3.5.2. Some valuation techniques of exotic options relevant to the book

Among all the options introduced in the last section, two aspects are especially important to our study of real options in large-scale engineering systems – compound options and path-dependency. This section will introduce the standard valuation techniques on compound options and path-dependency in the financial derivatives literature.

Compound options

For exotic options, one way to estimating value that is always available is Monte Carlo simulation, but European compound options can be valued analytically in terms of integrals of the bivariate normal distribution (Geske, 1979; Rubinstein, 1991). The value at time zero of a European call on a European call is

$$S_0 e^{-qT_2} M(a_1, b_1; \sqrt{T_1/T_2}) - X_2 e^{-rT_2} M(a_2, b_2; \sqrt{T_1/T_2}) - e^{-rT_1} X_1 N(a_2)$$

where

$$a_1 = \frac{\ln(S_0/S^*) + (r - q + \sigma^2/2)T_1}{\sigma\sqrt{T_1}}$$

$$a_2 = a_1 - \sigma\sqrt{T_1}$$

$$b_1 = \frac{\ln(S_0/X_2) + (r - q + \sigma^2/2)T_2}{\sigma\sqrt{T_2}}$$

$$b_2 = b_1 - \sigma\sqrt{T_2}$$

The function M is the bivariate normal distribution function and N is the normal distribution function (the same as the N function in Black-Scholes Formula). S_0 is the current stock price, q is the dividend yield, T_1 is the first exercise date, T_2 is the second exercise date, X_1 is the first strike price, X_2 is the second strike price, r is the risk-free interest rate, and S^* is the stock price at time T_1 for which the options price at time T_1 equals X_1. If the stock price is above S^* at time T_1, the first option will be exercised; otherwise, the option expires worthless.

With similar notation, the value of a European put on a call is

$$X_2 e^{-rT_2} M(-a_2, b_2; -\sqrt{T_1/T_2}) - S_0 e^{-qT_2} M(-a_1, b_1; -\sqrt{T_1/T_2}) + e^{-rT_1} X_1 N(-a_2)$$

the value of a European call on a put is

$$X_2 e^{-rT_2} M(-a_2, -b_2; \sqrt{T_1/T_2}) - S_0 e^{-qT_2} M(-a_1, -b_1; \sqrt{T_1/T_2}) - e^{-rT_1} X_1 N(-a_2)$$

the value of a European put on a put is

$$S_0 e^{-qT_2} M(a_1, -b_1; -\sqrt{T_1/T_2}) - X_2 e^{-rT_2} M(a_2, -b_2; -\sqrt{T_1/T_2}) + e^{-rT_1} X_1 N(a_2)$$

Path-dependent options

If the payoff of a derivative depends not only on the final value of the underlying asset, but also on the path followed by the price of the underlying asset, for example, several of the exotic options introduced in the previous section such as Asian options, lookback options, barrier options, then such derivative presents path-dependent features.

One approach that can always be tried to value path-dependent options when analytic results are not available is Monte Carlo simulation. A sample value can be calculated by drawing a realization of the path for the underlying asset. An estimate of the value of the derivative is obtained by the mean of a great number of realization paths. The main problem with the Monte Carlo simulation is that the computational cost to achieve the required level of accuracy can be prohibitively high, since the convergence rate of standard deviation is the square root of number of simulations. Another big problem with the Monte Carlo simulation is how to handle American options. Monte Carlo simulation is a forward looking algorithm while American options valuation needs backward deduction for optimal decision-making. Combining forward and backward algorithms together can be computationally too expensive to handle. A third problem with Monte Carlo simulation is that we can get the value but cannot understand the major drivers deciding the value of the derivative (like analytical solutions can provide). Despite the above mentioned problems with Monte Carlo simulation, Monte Carlo simulation is still probably the most widely applicable method to value path-dependent derivatives.

In many cases, we can also use binomial trees to cope with path-dependent options. For example, the shout option introduced in the previous section is a path-dependent derivative. We can value it by constructing a binomial tree for the underlying asset in the usual way. When we roll back the tree, we calculate the value of "shouting" and that of "not shouting" on each node. The value of the derivative on each node is the greater of the two. Hull and White [1993] suggested an extended binomial tree procedure to value path-dependent options. They gave examples using this procedure to value lookback options and barrier options.

For some path-dependent derivatives, we can also get analytic solution approximately. For example, we have approximate analytic valuation for Asian options. If the underlying asset S is assumed to follow a lognormal distribution and $S_{average}$ is taken to be a geometric average, $S_{average}$ is also lognormally distributed. And $S_{average}$ can be treated like a usual stock price and Asian options can be valued. However, Asian options are not defined in terms of geometric averages, but in arithmetic averages. Exact analytic pricing formulas are not available because the arithmetic average of a set of lognormal distribution does not have analytically tractable properties. But the distribution of arithmetic averages of a set of lognormal distribution is approximately lognormal and this leads to a good approximation for valuation of Asian options. We calculate the first two moments of the distribution of the arithmetic average and assume the distribution is lognormal with the two moments, then we can use standard valuation formulas.

Chapter 4 Real Options

"The classic way to value businesses is to compute the discounted present value of their future cash payouts. Not good enough, says Michael Mauboussin, the chief U.S. investment strategist at Credit Suisse First Boston. You should also throw in something for the company's 'real options'". (Schoenberger, 2000)

An article published in McKinsey Quarterly argued, "Real Options are especially valuable for projects that involve both a high level of uncertainty and opportunities to dispel it as new information becomes available". (Leslie and Michaels, 1997)

MIT professor Stewart Myers (1984) first coined the term "real options":

> "Strategic planning needs finance. Present value calculations are needed as a check on strategic analysis and vice versa. However, standard discounted cash flow techniques will tend to understate the option value attached to growing profitable lines of business. Corporate finance theory requires extension to deal with real options." (pp. 136)

Not until recently, the distinction between real options "in" and "on" projects has been drawn. Real options come in two basic flavors: those that are "on" systems and treat the technology as a black box, and those that are "in" systems and provide the flexibility and the option through the details of the design (de Neufville et al, 2004). This chapter will discuss real options "on" and "in" projects and some implications for the real options method.

4.1. Definition of Real Options

Before we start a detailed discussion, we need to define "real options" clearly. A real option is a right, but not obligation, to do something for a certain cost within or at a specific period of time. Compare the definition of real options with that of financial options on page 58: a financial option is restricted to buying or selling an underlying asset, and a real option refers more broadly to "do something", while the other aspects of financial and real options are similar.

The valuation of real options provides important insights into the value of opportunity or flexibility. Applying real options methodology, people can actively manage risks and uncertainties, not merely passively perceive the value of flexibility vaguely as before. People can systematically identify and establish options into a project, increasing the value of the project, appreciating the value of the project wholly, and taking advantage of upside potentials while avoiding downside risks.

Appreciating that a project is like a financial call option (or a put option) can help people recognize the crucial role that uncertainty plays in the investment decisions. For a financial call option, the more volatile the price of the stock in which the option is established, the more valuable is the option and the greater incentive to keep the option open. This is true because of the asymmetry in the option - the higher the price rises, the higher the payoff is; however, if the stock price falls, one can lose only what was the price of the option at the time of purchase.

The same goes for project investment decisions. The greater the uncertainty of a project, the greater the value of the opportunity and the greater incentive to wait and keep the opportunity alive rather that exercise it immediately. Of course, the traditional NPV method also considers uncertainty by way of the choice of discount rate. But in real options thinking, uncertainty is far more important and fundamental.

In addition to understanding the role of uncertainty, real options thinking helps companies to think systematically and actively to obtain options by their technological knowledge, reputation, managerial resources, market position, and possible scale. People need to understand options and get opportunities in hand first.

With some data, the real options approach can add quantitative rigor to the valuation of the flexibility. Flexibility comes with cost. Using quantitative real options valuation, people can calculate the net value of an option, i.e., the value of an option minus cost, given a certain amount of investment budget. With binomial and simulation valuation, moreover, people can get the possibility distribution of a project's payoffs with/without options. In this way, people can have a more holistic understanding of the project than if only the expected value of payoffs is given as in the traditional NPV method.

4.2. Real Options "on" Projects

Existing real options studies are mostly on real options "on" projects. There has been tremendous progress in the understanding of real options "on" projects since around 1995. This book stresses the difference between real options "on" and "in" projects. What is called real options "on" projects in this book are simply referred to as real options in the standard literature. We would like to contrast those two kinds of real options and stress the focus and contribution of this book on real options "in" projects.

4.2.1. What are Real Options "on" projects?

An opportunity is like a call option because the company has the right, not the obligation, to invest in a project. In the case of real options "on" projects, the company treats the engineering design of the project as a black box and values the black box. It is possible to find a call option sufficiently similar to the investment opportunity. The value of the option would tell us something about the value of the opportunity. Although most projects are unique and the likelihood of finding a similar option on the market is low, people can reliably find a similar option by constructing one.

Before the formalization of the options theory, people had intuitively knew the benefit of options, such as the ancient Chinese proverbs " a cunning rabbit has three caves" and "never put all the eggs in one basket". With the development of options theory, people can now estimate the value of opportunities more precisely, which enables them to compare the value of an option with its cost. A more scientific decision can be reached - people shouldn't spend more for an option than it's worth.

Example 1: Petroleum chemical company

Paraphrasing an example from Amram and Kulatilaka (1999), a petroleum chemical company might begin to invest in a new capacity, but is worried about the size of the market opportunity and whether the manufacturing process could meet the government regulations regarding environmental protection. Traditional Net Present Value (NPV) analysis suggested that the project should not be pursued. Real options "on" projects analysis, however, valued the exit option held by the company – the option that the company could walk away if there were bad news about the market or the government regulation. Although there would be a loss of initial investment if the project were cancelled, including the exit option, the project value increased and the company began to construct new capacity.

Example 2: Oil exploring

Paraphrasing an example from Leslie and Michaels (1997), a North Sea oil company accumulated a portfolio of license blocks – five-year rights to explore and produce oil and gas. The development was unsuccessful and left it with unwanted blocks that were consuming cash. The company decided to sell the blocks initially. During the divestment program, it was suggested, however, instead of calculating what the block would be worth if the company started developing them immediately, the company should value its opportunity as an option to develop if, at sometime in the future, recoverable reserves could be increased through new technologies. A simple financial model was developed to show how to price the blocks at their option value over 5 years, incorporating uncertainty about the size of the reserve, the oil prices, and room for flexible response to the outcome. The managers reevaluated the company's portfolio, and instead of letting blocks go, they held on to those with high option value and sold the rest at the revised values.

4.2.2. Comparison of Real options "on" projects Method and Traditional NPV Method

Often, although the NPV proves to be negative, the management team decides to go ahead anyway; or the NPV is positive, but intuition warns people not to proceed. It is not the intuition that is wrong, but the time-honored NPV decision-making tool. As a practical matter, many managers seem to understand there is something wrong with the simple NPV rules, i.e., there is a value to waiting for more information and this value is not reflected in the standard NPV calculation.

Traditional NPV valuation tools ignore an important value of a project - the value of flexibility. The traditional NPV method assumes, if an investment is irreversible, the investment is now-or-never, or in other words, if the company does not make the investment now, it will lose the opportunity forever. The traditional NPV method does not take into account an important reality: business decisions in many industries and situations can be implemented flexibly through deferral, abandonment, expansion, or in a series of stages that in effect constitute real options "on" projects.

See the following example based on Prof. de Neufville's class notes (2002): Suppose a project can be started for $100, and $1100 more will be required to complete. We must decide whether or not to continue after observing the initial result. And the commercial feasibility is decided by the initial result and the market condition then. Our final objective is to license the technology to a bidder who offers the highest price. The revenue estimate is shown in Table 4-1.

Table 4-1 Revenue Estimate for Technology Development

Revenue	Chance
License for $2000	50%
License for $100	50%

Assuming the discount rate is 10%, the question is: do we fund the project?

Table 4-2 shows traditional discounted cash flow (DCF) and net present value (NPV) valuation:

Table 4-2 NPV Valuation of Technology Development

	Year 0	Year 1	Year 3
Initial cost	-$100		
Development		-$1100	
License revenues			0.5x$2000 + 0.5x$100
Present Value	-$100	-$1000	$868

The traditional NPV valuation is -$232, so the project should be rejected.

But if we employ Real Options thinking, we understand that we have the option to develop only if the $2000 license is expected. Now the analysis is as Table 4-3. And the new NPV is $226, so we should accept the project[1]. Thinking of options is always natural and intuitive for managers even without the formal option valuation tools. However, we were not able to valuate options rigorously before we had option valuation models. Now, with those option valuation tools, options thinking can be transitioned from the state of qualitative intuition to the state of quantitative rigor.

Table 4-3 Approximate Options Valuation of Technology Development

	Year 0	Year 1	Year 2
Initial cost	-$100		
Development		0.5×$1100	
License revenues			0.5×$2000 + 0.5×$0
Present Value	-$100	-$500	$826

In addition, there is another key difference between Real options "on" projects valuation and NPV. NPV needs an appropriate discount rate to bring the future cash flows back into present dollars, while real options "on" projects models are attractive because they eliminate the need to resolve this issue. The Black Scholes Formula (Equation 3-10) shows that options pricing does not require a discount rate. The question regarding how to decide an appropriate discount rates generates a lot of debate. There is no consensus or natural way to get an appropriate discount rate for a project. The Real Options approach can circumvent the ambiguity of discount rate choice.

4.2.3. Types of Real options "on" projects

---

[1] Note this $226 is not the options value. It is only an approximation of the options value because it is not a risk-neutral valuation.

Some options occur naturally (e.g., to defer, contract, shut down or abandon), while others may be planned and built-in with extra cost (e.g. to expand growth options, to default when investment is staged sequentially, or to switch between alternative inputs or outputs). Table 4-4 describes briefly the most common categories of real options "on" projects.

Table 4-4 Types of Real options "on" projects

Category	Description	Important In
Option to defer	Management holds a lease on (or an option to buy) valuable land or resources. It can wait (x years) to see if output prices justify constructing a building or plant, or developing a field.	All natural resource extraction industries; real estate development; farming; paper products
Time to build option (staged investment)	Staging investment as a series of outlays creates the option to abandon the enterprise in midstream if new information is unfavorable. Each stage can be viewed as an option on the value of subsequent stages, and valued as a compound option.	All R&D intensive industries, especially pharmaceuticals; *long-development capital-intensive projects, e.g., large-scale construction or energy-generating plants*; start-up ventures
Scaling Option (e.g., to expand; to contract; to shut down or restart)	If market conditions are more favorable than expected, the firm can expand the scale of production or accelerate resource utilization. Conversely, if conditions are less favorable than expected, it can reduce the scale of operations. In extreme cases, production may halt or start up again.	Natural resource industries such as mine operations; facilities planning and construction in cyclical industries; fashion apparel; consumer goods; commercial real estate.
Option to abandon	If market conditions decline severely, management can abandon current operations permanently and realize the resale value of capital equipment and other assets in secondhand markets.	Capital-intensive industries, such as airlines and railroads; financial services; new product introductions in uncertain markets.

Option to switch (e.g., outputs or inputs)	If price or demand change, management can change the output mix of the facility ("product flexibility"). Alternatively, the same outputs can be produced using different types of inputs ("process flexibility")	Output shifts: Any goods sought in small batches or subject to volatile demand, e.g., consumer electronics; toys; specialty paper; machine parts; autos; Input shifts: All feedstock-dependent facilities, e.g., oil; electric power; chemicals; crop switching; sourcing
Growth option	As early investment (e.g., R&D, lease on undeveloped land or oil reserves, strategic acquisition, information network/infrastructure) is a prerequisite or link in a chain or interrelated projects, opening up future growth opportunities (e.g., new generation product or process, oil reserves, access to new market, strengthening of core capabilities). Like interproject compound options.	All infrastructure-based or strategic industries, especially high-tech, R&D, or industries with multiple product generations or applications (e.g. computers, pharmaceuticals); multinational operations; strategic acquisitions.
Multiple interacting options	Real-life projects often involve a "collection" of various options, both upward-potential enhancing calls and downward-protection put options present in combination. Their combined option value may differ from the sum of separate option values, i.e., they interact. They may also interact with financial flexibility options.	Real-life projects in most industries discussed above.

(Source: Lenos Trigeorgis, 1993. Real Options and Interactions with Financial Flexibility. Financial Management. Autumn.)

4.2.4. Framework of Real options "on" projects Valuation

Before using real options "on" projects to evaluate a project, we first need to understand clearly what decisions need to be made and check if it is advantageous to use this approach over the traditional NPV method. If so, the valuation can be divided into six steps, as shown in Figure 4-1:

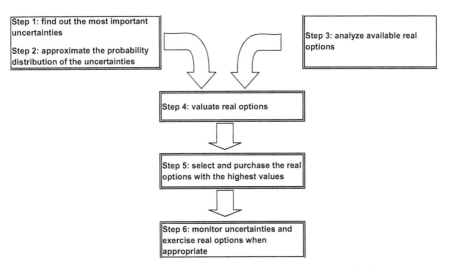

Step 1: find out the most important uncertainties

Step 2: approximate the probability distribution of the uncertainties

Step 3: analyze available real options

Step 4: valuate real options

Step 5: select and purchase the real options with the highest values

Step 6: monitor uncertainties and exercise real options when appropriate

Figure 4-1 Framework of Real options "on" projects Method

As a first step, most important drivers and uncertainties of the project should be found out. Usually uncertainties include market risk (such as the market demand, price of the product, economic cycle), technical risk (such as if the project can be finished on time, if the project can achieve its technical objectives).

The second step, an approximate probability distribution should be assigned to each uncertainty. In many cases, a lognormal distribution is used for market risk. If there are other project-specific risks (private risks) associated with the project, their probability distributions should be studied case by case.

In the third step, interesting options should be identified. Possible options practical to the project studied can be identified with reference to Table 4-4 for the types of real options "on" projects.

The fourth step, appropriate method among Black-Scholes formula, binomial model, and simulation is identified and applied to obtain the value of the options.

The fifth step, by comparing the value of the options and cost to obtain options, most worthwhile options are selected and purchased or implemented.

The sixth step, the key uncertainties are closely monitored, and the purchased or implemented real options will be exercised when appropriate.

Analysts and managers need to be careful of the false precision of the value of an option, because the value is established on many approximations and assumptions. This is why a sensitivity analysis is sometimes needed. Nevertheless, the mind-set to value the flexibility is one of the major gains of this thesis.

4.2.5. Real options "on" projects versus Financial Options

To use the methodology originally developed for financial options, there should be an appropriate underlying. For the most talked-about stock options, the underlying is the stock price. For real options "on" projects study, however, a generalized concept of underlying is needed. An underlying is the agent that determines the value of a project or an investment. An underlying can be assets, but it also can be other agents such as market size or utility price. To use Black-Scholes or the binomial model, the underlying should follow the Geometric Brownian motion like stock prices.

If the underlying is not following a geometric Brownian motion, the options thinking can still be applied. The reason is that the key for options thinking does not necessarily have to resemble financial options exactly. The essence is the right not the obligation for a property or project. If appropriate stochastic processes for underlying can be found, people can use mathematical deduction to get the valuation (like Black-Scholes formula to get option value based on the assumption of geometric Brownian motion), or can just use Monte Carlo simulation to get the option value.

There are a number of difficulties to apply real options "on" projects to fields beyond exchanges and over-the-counter markets where options were originally developed. The key problem is that often no efficient market exists for the object studied, so the most powerful characteristics of financial options theory are mostly gone. For financial options theory, the power originates from the assumption, close to reality, that stock prices in a an efficient market contain all the information.

Often, the real options "on" projects approach is hard to be applied because of the absence of justifiable value of volatility, the key variable for real options analysis. One of the ways to circumvent this problem is to use the hybrid real options "on" projects method (Neely and de Neufville, 2001) that uses decision analysis for the part of analysis where historical data is not sufficient, e.g., the R&D stage for a new product.

4.3. Real Options "in" Projects

Real options "in" projects is the latest extension of real options work into physical systems. See Figure 4-2. The concept is new and the methodology needs to be further developed. A lot of the following text is from a paper by the author and de Neufville (2005).

4.3.1. What are real options "in" projects?

Again we describe the two basic flavors of real options: those that are "on" systems and treat the technology as a black box, and those that are "in" systems, and provide the flexibility and the option through the details of the design (de Neufville et al, 2004). A simple example of a real option "in" a system is a spare tire on a car: it gives the driver the "right, but not the obligation" to change a tire at any time, but this right will only rationally be used when the car has a flat tire.

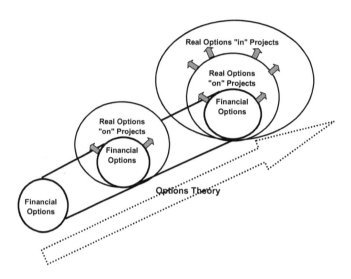

Figure 4-2 Development of Options Theory

Real options "in" projects are of special interest to the study of engineering systems. Large-scale engineering projects share three major features. As Roos (2004) has indicated, they

- Last a long time, which means they need to be designed with the demands of a distant future in mind;
- Often exhibit economies of scale, which motivates particularly large construction;
- Yet have highly uncertain future requirements, since forecasts of the distant future are typically wrong.

This context defines the desirability of creating designs that can be easily adjusted over time to meet the actual needs as they develop. System leaders need to build "real options" into their designs. Engineers increasingly recognize the great value of real options in addressing intrinsic uncertainties facing large-scale engineering systems and, more importantly, are learning to manage the uncertainties proactively (de Neufville et al, 2004).

Note the difference between real options "in" projects and the engineering concept of "redundancy". Both real options "in" projects and redundancy refer to some components should not have been designed if the design were optimized given the assumption that things are not going to change. Redundancy refers to duplication of design elements serving to increase the reliability of the system in case of components failures, while real options "in" projects may not serve the same functions as some currently existing components (though such real options may not prove necessary given the current situation).

Real options "in" systems are those that are most interesting to designers, and are the focus of this book. Following are several examples of real options "in" projects for engineering systems.

Example 1: "Bridge in bridge"

The design of the original bridge over the Tagus River at Lisbon provides a good example of a real option "in" a major infrastructure system. In that case, the original designers built the bridge stronger than originally needed, strong enough so that it could carry a second level, in case that was ever desired. The Portuguese government exercised the option in the mid 1990s, building on a second deck for a suburban railroad line (Gesner and Jardim, 1998).

Example 2: Satellite systems

In the late 1980s, Motorola and Qualcomm planned the Iridium and Globalstar systems to serve their best estimates of the future demand for space-based telephone services. Their forecasts were wrong by an order of magnitude (in particular because land-based cell phones became the dominant technology). The companies were unable to adjust their systems to the actual situation as it developed and lost almost all their investments -

- 5 and 3.5 billion dollars respectively. However, if the companies had designed evolutionary configurations that had the capability to expand capacity, it would have been possible both to increase the expected value of the system by around 25%, as well as to cut the maximum losses by about 60% (de Weck et al, 2004). Such evolutionary configurations can be realized by designing real options for the room of future capacity expansion. For example, a smaller system with smaller capacity can be established first. For a smaller system, there could be fewer satellites with a higher orbit. One possible real option is to carry extra fuel on each satellite. When demand proves big, the satellites can move to lower orbits with the existing orbital maneuvering system (OMS). With additional satellites launched to lower orbits, a bigger system is accomplished to serve the big demand. The extra fuel carried in the satellites are real options. They can be exercised when the circumstances turn favorable. There is cost to acquire such real options – the cost of designing larger tanks and launching extra fuel. Decision makers have the right to exercise the options, but not the obligation – they can leave the extra fuel on board. The key point is that we have to put some provisions in the initial system that enable the system to respond to some uncertainties.

Example 3: Parking garage design

This example is based on a technical note by de Neufville, Sholtes, and Wang (2005), and more details of this example follows in this section. A car parking garage for a commercial center is planned in a region that is growing as population expands. Economic analysis recognizes that actual demand is uncertain, given the long time horizon. If the owners design a big parking garage, there is possibility that the demand is smaller and the cost of a big garage cannot be recovered; however, if the owners design a small parking garage, they may miss the opportunity if the demand grows rapidly. To deal with this dilemma, the owners can design real options into the design by strengthening the footings and columns of the original building so that they can add additional levels of parking easily. This premium is the price to get the real option for future expansion, a right but not an obligation to do so.

4.3.2. Comparison of real options "on" and "in" projects

Real options "on" projects are mostly concerned with valuation of investment opportunities, while real options "in" projects are mostly concerned with design of flexibility. The classic cases of real options "on" projects are on valuation of oil fields, mines, and pharmaceutical research projects, where the key question is to value such projects as a whole and decide if it is worth to invest in the projects. The examples of real options "in" projects are extra fuel on satellites, strengthened footings and columns of a multi-level parking garage, or the "bridge in bridge".

Real options "on" projects are mostly concerned with an accurate value to assist sound investment decisions, while real options "in" projects are mostly concerned with "go" or "no go" decisions and an exact value is less important. For real options "on" projects, analysts need to get the value of options, but for real options "in" projects, analysts do not have to provide the exact value of the options and simply provide what real options (flexibility) to design into the physical systems.

Real options "on" projects are relatively easy to define (Table 4-4 lists seven kinds of most common real options "on" projects and this covers most of the cases), while real options "in" projects are difficult to define in physical systems. For an engineering system, there are a great number of design variables, and each design variable can lead to real options "in" projects. It is hard to find out where the flexibility can be and where is the most worthy place to design real options "in" project. Identification of options is an important issue for real options "in" projects.

Real options "on" projects do not require knowledge on technological issues, and interdependency/path-dependency is not frequently an issue. However, real options "in" projects need careful consideration of technological issues. Complex technological constraints often lead to complex interdependency/path-dependency among projects. We will further discuss this issue in the book.

Real options "on" projects	Real options "in" projects
Value opportunities	Design flexibility
Valuation important	Decision important (go or no go)
Relatively easy to define	Difficult to define
Interdependency/Path-dependency less an issue	Interdependency/Path-dependency an important issue

Table 4-5 Comparison between real options "on" and "in" projects

Table 4-5 summarizes the comparison between real options "on" and "in" projects.

4.3.3. Difficulties facing the analysis of real options "in" projects – Options Identification and Path-dependency

Besides knowledge of technology, there are more difficulties facing the analysis of real options "in" projects:

1. Financial options are well-defined contracts that are traded and that need to be valued individually. But real options "in" projects are fuzzy, complex, and interdependent: To what extent is there a predetermined exercise price? What is the time to expire? Moreover, it is not obvious that is is useful to value every element that provides flexibility.

2. Real options "in" projects are likely to be path-dependent – they depend not only on the final value of the underlying, but also on the path followed by the evolution of the underlying. For example, the capacity of a thermal power system at some future date may depend on the evolutionary path of electricity use. If the

demands on the system have been high in preceding periods, the electric utility may have been forced to expand to meet that need, as it might not have done if the demand had been low. Real options "in" projects may thus differ fundamentally from stock options, whose current value only depends on the price at that time except in some of the exotic options. The evolutionary path of a stock price does not matter. Its option value is path-independent. This is not true for many real options.

Real options "in" projects are also likely to be highly interdependent, compound options. Their interactions need to be studied carefully as they may have major consequences for important decisions about the design of the engineering system. The associated interdependency rapidly increases the complexity and size of the computational burden.

To develop a method for building real options "in" physical projects, the book offers suggestions for addressing the above difficulties:

1. It proposes to identify candidate real options "in" projects by screening and simulation models. This is important because, in an interdependent system, it may not be obvious where flexibility in the system may be most valuable. The book focuses on developing the most appropriate designs of flexibility and building up suitable contingency plans for dealing with future uncertainties.
2. To simplify highly complicated path-dependent problem, the book proposes to divide the decision time horizon into a small number of periods, then solve the path-dependent problem by a timing model using stochastic mixed-integer programming. This process also deals with compound options difficulty mentioned above as well.

4.3.4. A Case Example on analysis of Real Options "in" Projects – Parking Garage

The case example of the design of a parking garage is inspired and extrapolated from the Bluewater development in England (http://www.bluewater.co.uk/). This case example has been developed into a technical note by de Neufville, Scholtes, and the author of the book (2005). The remainder of this section is cited from the technical note.

The case deals with a multi-level car park for a commercial center in a region that is growing as population expands. The basic data are that:
- The deterministic point forecast is that demand on opening day is for 750 spaces, and rises exponentially at the rate of 750 spaces per decade;
- Average annual revenue for each space used is $10,000, and the average annual operating cost for each space available (often more than the spaces used) is $2,000;
- The lease of the land costs $3.6 Million annually;
- The construction will cost $16,000 per space for pre-cast construction, with a 10% increase for every level above the ground level;
- The site is large enough to accommodate 200 cars per level; and
- The discount rate is taken to be 12%.

Additionally, economic analysis needs to recognize that actual demand is uncertain, given the long time horizon. The case assumes that future demand could be 50% off the projection, either way, and that the annual volatility for growth is 15%.

Real options (flexibility): The owners can design the footings and columns of the original building so that they can add additional levels of parking easily, as was the case for the Bluewater development. The case assumes that doing so adds 5% to the total initial construction cost. This premium is the price to get the real option for future expansion, a right but not an obligation to do so.

We use an Excel spreadsheet simulation model to analyze the real options embedded in the design of the parking garage. The real options analysis using spreadsheets involves 3 steps:

1. Set up the spreadsheet representing the most likely projections of future costs and revenues of the project, and calculate its standard engineering economic value. The design that maximizes the NPV is the base case against which flexible solutions are compared, so as to derive the value of these alternative designs.

2. Explore the implications of uncertainty by simulating possible scenarios. Each scenario leads to a different NPV, and the collection of scenarios provides both an "expected net present value" (ENPV) and the distribution of possible outcomes for a project. These are usefully plotted as cumulative distribution functions that document the Value at Risk (VaR), that is, the probabilities that worse cases could occur. This documentation motivates the search for the flexibility, for the real options, that will enable the managers of the infrastructure to avoid these losses.

3. Analyze the effects of various ways to provide real options by changing the costs and revenues to reflect these design alternatives. The difference between the resulting best ENPV and that of the base case is the value of real options. Moreover, the VaR curve for the flexible design intuitively explains how real options allow system operators to avoid downside losses and take advantage of upside opportunities. This information can be a key factor in decisions about the design of major projects.

Following the three steps, we analyze the case example:

Step 1: Table 4-6 illustrates the basic spreadsheet for calculating the NPV of the parking garage, assuming that the demand for spaces grows as projected. Note that the project cannot benefit from addition demand when it exceeds the capacity of the facility.

The designer can use the spreadsheet to calculate the NPV for any number of levels for the car park (Figure 4-3) and thus determine the size that maximizes NPV. The optimal design for this base case, that unrealistically assumes that demand is known in advance, is to build 6 floors. Its apparent NPV is $6.24 million. This estimate is however wrong: actual demand will vary from the deterministic forecast, so that the ENPV of this design will also be different, as Step 2 documents.

Step 2: Recognizes the uncertainty in the forecast demand by simulating possible scenarios, S. This example analysis ran 2000 scenarios according to random draws from the stochastic process for the uncertain demand, which took about 1 minute on a standard PC. Each scenario implies a different NPV. The set of scenarios thus represents the probability distribution of the NPV that might occur. As Figure 4-3 indicates, the actual expected NPV for the deterministic design under probabilistic scenarios is less than that estimated from a deterministic analysis. It is only $2.87 million. In fact, the smaller 5-level design provides greater expected NPV ($ 2.94 million) since it lessens the possibility of big losses from overbuilding capacity that might not be used.

The analysis considering uncertainty provides useful insights that should motivate designers and decision-makers to use real options "in" projects. It shows that:
- Uncertainty can lead to asymmetric returns. In this case, although the case assumes that the chance of higher and lower demands are equal, the upside value of the project is limited (because the fixed capacity cannot take advantage of higher demands) while the downside risks are substantial and can lead to great losses.
- The actual expected value of a project P over all the scenarios in general is not equal to the value of a project for an average scenario, as Figure 4-3 indicates. This is the "Flaw of Averages" or Jensen's Inequality (Savage, 2000):

$$EV\,P(S) \neq P\,[EV(S)]$$

- The cumulative distribution gives the Value at Risk (VaR). It shows the probability that an NPV might be less or equal to a threshold. Thus Figure 4-4 shows that there is about 10% chance that the losses from the 5-level parking garage would exceed $4 million

Category	Type	Units	Year					
			0	1	2	3	...	20
Demand		Spaces		750	893	1,015	...	1,696
Capacity	Initial			1,200	1,200	1,200		1,200
Revenue				7.50	8.93	10.15		12.00
Cost	Initial	$ Millions	22.74					
	Annual		3.60	6.00	6.00	6.00		6.00
Cash Flow	Actual		- 26.34	1.50	2.93	4.15		6.00
NPV			6.24					

Table 4-6 Spreadsheet for Design with Deterministic Point Forecast of Demand

(Case of 6 level garage)

Step 3: Explore ways to limit the downside risk and take advantage of upside potential. For example, designers can reduce losses by creating smaller designs that lower the chance that demand will not fill the facility. In this case, the smaller design eliminates the chance of really big losses, but at the cost of never making any substantial profit. Thus, as is frequently the case, simply providing good insurance against losses is not sufficient to make a project attractive.

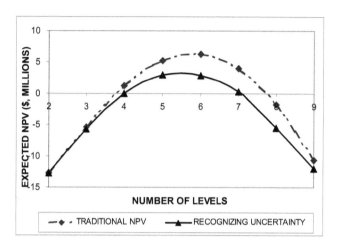

Figure 4-3 Expected Net Present Value for Designs with Different Number of Levels

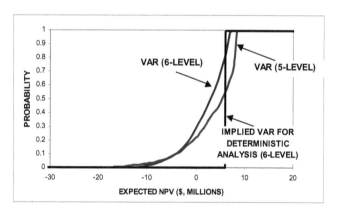

Figure 4-4 Value at Risk for 6 level design recognizing demand uncertainty, compared to deterministic estimate

Designers can take advantage of possible growth by building expansion options into the design. As was done in the parking structures for the Bluewater development, this case considered the possibility of making the columns big enough to support additional levels, should demand justify expansion of the parking garage in later years. Table 4-7 shows the spreadsheet to explore this expansion option, with appropriate modifications in bold type. It incorporates additional rows for "Extra capacity", and "Expansion cost". For this case, the decision to construct an extra floor or 200 spaces was made if the capacity was less than the demand for two consecutive years. Other criteria and rules could be programmed in.

The graphical interpretation is that the designer shifts the VaR curve to the right by reducing the extent of the lower tail into losses, and pushing the upper tail into gains. Figure 4-5 shows the joint VaR of building small with the option to expand if demand is favorable. The initial design of only 4 levels greatly decreases the maximum loss (from 24.68 to 12.62 million). The capability to add capacity increases both the maximum value of the project (from 13.78 to 14.80 million) and its expected value. The estimated value of the options embedded in the flexible design is the difference between the expected value with the options ($5.12 million) and the base case design defined in the standard deterministic way (2.87 million), that is $2.25 million in this case.

Category	Type	Units	Year					
			0	1	2	3	...	20
Demand				1055	1141	1234	...	1598
Capacity	Initial	Spaces		800	800	**1,000**		**1,800**
	Added					**200**	**200**	
Revenue				8.00	8.00	10.00		15.98
Cost	Initial	$ Millions	14.48					
	Later				**4.26**	**4.68**		
	Annual		3.60	5.20	5.20	**5.60**		7.20
Cash Flow	Actual		-18.08	2.80	-1.46	-0.28		8.78
NPV			7.57					

Table 4-7 Spreadsheet for Design with One Scenario of Demand and Option to Expand

(Case of 4 level garage)

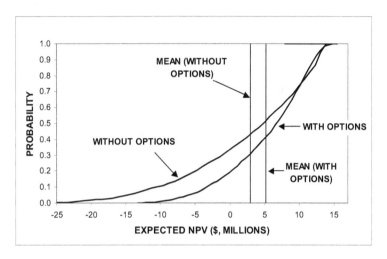

Figure 4-5 Option to Expand adds significant value and improves profile of VaR

Perspective	Step of Analysis	Simulation Used?	Has Option?	Design Levels	ENPV $, Millions
Deterministic	1	No	No	6	2.87
Recognizing Uncertainty	2	Yes	No	5	2.94
Incorporating Real options "in" projects	3	Yes	Yes	4, with strong columns	5.12

Table 4-8 Comparison of 3 Steps of Analysis

The flexibility provided by building small initially with the option to expand has several advantages beyond increasing the expected value of the project. The spreadsheet approach to the real options analysis generates the data that bring out these features, as the financial approaches do not. Table 4-9 presents this information and provides a multi-faceted analysis and justification of the flexible approach to design. In this case the analysis documents that the flexible design of the multi-level garage:

- Reduces the maximum possible loss, that is the Value at Risk;
- Increases the maximum possible and the expected gain;
- While maintaining the initial investment costs low.

Metric	Design		Comparison
$, Millions	No Real Options	With Real Options	
Initial Investment	22.74	14.48	Real Options Better
Expected NPV	2.87	5.12	Real Options Better
Minimum NPV	-24.68	-12.62	Real Options Better
Maximum NPV	13.78	14.80	Real Options Better

Table 4-9 Performance Improvements achieved with Flexible Design

(Maxima and Minima of simulation taken at 0.05 and 99.5 percentile)

4.4. Possible valuation techniques for real options

In this section, we will examine the applicability of the three most important options valuation techniques to real options. The three techniques are the Black-Scholes formula, simulation, and the binomial lattice method.

4.4.1. Black-Scholes Formula

As derived in Section 3.3.1. , the Black-Scholes formula for the prices at time zero of a European call option on a non-dividend-paying stock[1]:

$$c = S_0 N(d_1) - X e^{-rT} N(d_2)$$

where

$$d_1 = \frac{\ln(S_0 / X) + (r + \sigma^2 / 2)T}{\sigma \sqrt{T}}$$

$$d_2 = \frac{\ln(S_0 / X) + (r - \sigma^2 / 2)T}{\sigma \sqrt{T}} = d_1 - \sigma \sqrt{T}$$

and $N(x)$ is the cumulative probability distribution function for a variable that that is normally distributed with a mean of zero and a standard deviation of 1.0.

The formula is the result of solving a Partial Differential Equation (PDE), seemingly opaque and incomprehensible to those not familiar with financial mathematics or physics. Moreover, lacking an understanding of the underlying assumptions for the Black-Scholes formula, it is very easy to apply the formula blindly and lead to useless and misleading precise value of "value of real options". The major assumptions underlying Black-Scholes approach are:

1. There are prices for the asset;
2. Efficient market for the asset with no riskless arbitrage opportunities, and some special conditions for the market:
   o the short selling of securities has no limitation,

---

[1] Similar formulas can be derived for European put options, and European call or put options with dividend paying.

- o no transaction costs or taxes,
- o all securities are perfectly divisible,
- o security trading is continuous,
- o the risk free rate of interest is constant and the same for all securities;
3. The price of the underlying asset follows Geometric Brownian Motion with μ and σ constant.

For the big picture of this study on real options and for simplicity, we can regard that the special conditions for the market in the above point 2 are approximately satisfied or, even if it is not perfectly satisfied, they are secondary in comparison to the three major points and have a much smaller impact on the valuation. Now let us examine the three most important assumptions:

1. The price assumption for Black-Scholes approach is not discussed in finance literature, since you must have prices then have a financial market, stocks, derivatives, and theories. But for real options, it is sometimes not the case that the analyst has a market price for the object studied. However, there may be market prices for the final products and the dynamics for the prices may be well understood, for example, oil field or copper mine. For some other cases, it may not be easy to decide the dynamics of market price for the products of a system, for example, computers. For still some other cases, it may not even be possible to decide market price for the product of a system, for example, national defense, space exploration.

2. The no arbitrage condition is often hard to satisfy for real options. If people can construct a replicating portfolio PERFECTLY to match the payoff of the real options under all possible situation, then an arbitrageur can take advantage of the mismatch of the price between the portfolio and the real options, and earn profit RISKLESSLY. If the price for the real options is too high, arbitrageurs could sell the real options and buy the replicating portfolio to earn riskless profit; else if the price for the real options is too low, arbitrageurs could sell the replicating portfolio

and buy the real options to earn a riskless profit. Since such activities of arbitrageurs will change the demand and supply of the real options on the market, and finally drive the price of the real options to equal that of the replicating portfolio. Such "no arbitrage" is usually hard to prove valid for a real option. The payoff of a stock option can be perfectly matched by a portfolio of stocks and loan, but how can we match a real option? Sometimes, we can assume a reasonable approximation for the replicating portfolio, for example, purchase of an oil field with the option to postpone development can be replicated by a portfolio of long position in oil futures and borrow money for the position. Often, however, it is not possible to find replicating portfolio for real options. For example, the real options of strengthened footings and columns in the parking garage cases, how to replicate the real options?

3. The Geometric Brownian motion assumption has the property that the price grows forever. For some underlying assets, it is an acceptable assumption, for example, the stock price because of continuous inflation and investment. For some underlying assets, however, the Geometric Brownian motion is not a best assumption. For example, the case study in this book is about river basin development with the purpose of power generation, the underlying for the case study is electricity price. Empirical evidence shows that Geometric Brownian motion is not the best model to describe the stochastic movement of electricity price, and MRPV process is a better model (Bodily and Buono, 2002). Constant μ and σ is needed for Black-Scholes approach even if Geometric Brownian motion assumption is validated. Fortunately, if μ or σ vary with respect to time, we have means to deal with such relaxation of assumption in finance theory.

With the above discussion of assumptions for Black-Scholes, we can conclude Black-Scholes approach may be valid for real options "on" projects, but it hardly works for real options "in" projects where replicating portfolios are almost impossible to define.

4.4.2. Simulation

Monte Carlo simulation does not require a myriad of assumptions as the Black-Scholes formula. If only we can specify the stochastic processes for the underlying uncertainties, and we can understand the function between the input uncertain variables and the output payoff, we can just let computers do the "brute force" work. Plausibly, simulation can obtain any valuation that Black-Scholes can get at any specified level of accuracy, and it can tackle problems with complex and non-standard payoffs that Black-Scholes cannot deal with. However, we have to understand several issues before using the Monte Carlo simulation:

1. We have to have sound stochastic models for the underlying uncertain variables, especially the parameters in the stochastic models. If we use the wrong model or wrong parameters, the simulation model can give erroneous results. If the analyst uses the common geometric Brownian motion blindly in the simulation without checking its validity in the special context, the results are not only useless but also misleading.

2. The computational cost could be expensive for simulation methods. To get the required accuracy, the convergence could be slow and be very time consuming. In this context, variance reduction procedures are very important in the application of simulation. The important variance reduction procedures include antithetic variable technique, control variate technique, importance sampling, stratified sampling, moment matching, quasi-random sequences, representative sampling through a tree. A brief introduction of applications of variance reduction procedures in finance can be found in Hull (1999).

3. Simulation is not a panacea; there are cases where simulation is impotent. The "Curse of Dimensionality" refers to the case where the number of samples per variable increase exponentially with the number of variables to maintain a given level of accuracy. If there are multiple sources of uncertainty, then it could be

computationally prohibitive to calculate the option value at the required accuracy. Also, simulation needs an analytic form of the exercise condition for the options. If there are no closed-form analytical exercise conditions, for example American options, the simulation technique could not work without special treatment. If the backward looking optimality criterion for American options is used, it excludes the possibility of straightforward use of implicitly forward looking simulation technique.

4. Simulation can only provide a value, but does not shed light on the intrinsic relationship between variables and does not provide insights into what and how are the key drivers for the valuation. Black-Scholes formula provides a closed-form analytic solution, which allowed people for the first time to understand the important role of volatility in options pricing, and allowed people to calculate sensitivity measures such as Greek letters easily to gain insights into hedging. Simulation provides much less such critical insights.

With the understanding of the issues and limitation of simulation technique, we can unleash the power of the simulation in valuation of real options because of its versatility and low requirement on assumptions.

4.4.3. Binomial Tree

The binomial tree is a dynamic programming algorithm. It is not necessary binomial, and it could be trinomial or more. But whatever multinomial is, the essence is the same as the binomial lattice method - they allow the recombination of states to decrease the computational burden. With the number of nodes grows at only one for each additional stage considered, we can improve the precision of binomial tree method to a very high level by dividing the life span of an option into more stages.

Binomial trees work with both risk-neutral valuation and actual valuation. Risk-neutral valuation uses risk-neutral probabilities and discounts at the risk-free interest rate; actual valuation uses actual probabilities and discounts at a risk-adjusted discount rate. When risk-neutral valuation is possible, i.e. no arbitrage condition holds, and applied correctly, risk-neutral valuation and actual valuation will give the same result. Black-Scholes approach is risk-neutral valuation; if it is applied in the case where the "no arbitrage" condition does not hold, Black-Scholes will simply render useless and misleading results. For real options, often the no arbitrage condition does not hold, we should not naively use the Black-Scholes formula, but we can still use binomial tree.

The tree structure actually can deal with more than Geometric Brownian Motion implied by standard binomial tree. We can establish different trees for different stochastic processes. The recombination structure of the binomial tree implies path-independency. If the new process has path-dependent features, we can break the recombination structure of the tree. Although with the recombination structure broken, the number of nodes increases exponentially rather than arithmetically when the number of periods increases, for a small number of stages, it is still maneuverable. The standard binomial tree implies path-independency, while many real options present path-dependent features. We need to break the recombination structure to deal with the path-dependency.

Finally, we should point out that Black-Scholes, simulation, and binomial tree techniques should give exactly the same answer for the same valuation problem. However, depending on the circumstances, some techniques may be more effective or accurate than the other. To summarize,
- Black-Scholes approach should be used with great care when applied to real options, we have to justify its assumptions;
- Simulation is very useful but we need to understand its limitation and apply variance reduction techniques;
- The binomial tree method is versatile and powerful, but keep in mind if path-dependency presents (as it is common for real options "in" projects), we have to break the recombination structure of the tree and limit the number of periods considered.

4.5. Some Implications of Real Options Method

With the real options methodology, we can value opportunities or design flexibility to enhance our ability to manage uncertainties proactively in a fast-changing environment. In this section, we present some implications of the real options method

4.5.1. Investment with Options Thinking

In an uncertain world, several kinds of strategic investments can be analyzed from a real option perspective:

Irreversible investments

Irreversible investment requires more careful analysis because, once the investment takes place, the investment cannot be recouped without a significant loss of value. With the real options analysis, it is understood that irreversible investments, for most of the cases, should be delayed until a significant amount of the uncertainty is resolved, or the investments should be broken into stages.

Flexibility investments

Flexibility investment builds options into the initial design. Flexible design allows a production line to be easily switched across products. The option to switch is part of the capital investment.

Insurance investments

Insurance investments reduce exposure to uncertainty. Investment in excess capacity ensures against demand surges, but with a cost or "insurance premium". Decision-makers using real options approach are able to value the flexibility and check to see whether the value exceeds the cost.

Platform investments

Platform investments create valuable follow-on contingent investment opportunities. Using Real Options approach, managers can create a portfolio of projects, maximizing the value of the portfolio, balancing the portfolio with high-risk-high-return and low-risk-low-return projects, and aligning the projects tightly with the corporate strategy.

Growth investments

Growth investments are made to obtain information that is otherwise unavailable. For example, oil exploration is a growth investment because it generates geological information.

4.5.2. Value of Real options Method in Different Situations

Real options valuation is important in situations with high uncertainty and people have many options when new information is received. If the uncertainty is low and the available practical options are few, the Real Options approach will not add much insight beyond the traditional NPV method. This is because the flexibility value is near zero. Please see Figure 4-6.

For the case of high uncertainty and abundant available options, the flexible strategy and the real options approach are most valuable. And the Real Options approach will provide a much better result than NPV method.

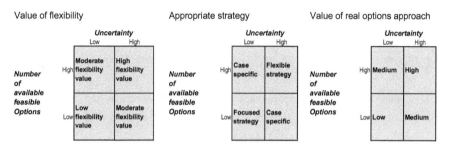

Figure 4-6: Applicability of Real Options Method

4.6. Attack and defense of real options method

Some people doubt the theory of real options. They believe the essence and beauty of financial options theory lies in arbitrage enforced pricing or contingent claims analysis. However, it is hard to see that arbitrage enforced pricing is relevant in many cases of real options. In many cases, using the real options method is hard to avoid the problem of decision on the risk adjusted discount rate and decision maker's subjective valuation of risk. This implies that real options analysis cannot obtain an objective valuation based on market observable prices, and people can maneuver the real options analysis. Everybody can reach a different result from his/her own real options analysis and there is no possibility to prove who is correct and who is wrong, because the subjective valuation of risk enters the analysis.

Despite all the doubts, real options theory is popular and developing fast... The great German philosopher Hegel said, "Whatever is reasonable is true, and whatever is true is reasonable." The author has an explanation on why real options theory is popular and highly useful.

4.6.1. Arbitrage-enforced pricing and real options

For arbitrage enforced pricing to work, we must understand how arbitrage opportunities are removed. The crux most relevant to real options lies in two points:

- There is some traded asset that has stochastic components that obey the same probability law and are perfectly correlated with the real options, and
- Arbitrageurs are able to short sell the real options[1].

If these two conditions and some other conditions are true, an arbitrageur can construct a portfolio to replicate the options perfectly and remove all risk. The arbitrageur then earns the risk-free rate since there is no risk involved. If the arbitrageur earns more than the risk-free rate, there is arbitrage opportunity, and the arbitrageurs' activities will eliminate such opportunities quickly. If arbitrage-enforced pricing works, we can prove that there is a market price of risk, which is the same for all derivatives that are dependent on the same risk at the same time. With the market price of risk, we can link the risk-free rate and risk-adjusted discount rate and helps us move from a world with risk preference to a risk neutral world. The valuation obtained from the risk neutral world is valid in the worlds with risk preference. With the validity of risk neutral valuation, we can obtain an objective value of options independent of individual risk preference – a very difficult part of analysis.

For real options, it is hard to find a traded asset that has stochastic components perfectly correlated with the real options. If it is possible for some real options "on" projects, it is almost never the case for real options "in" projects. Moreover, many real options are large-scale projects, so that short selling of the real options is not realistic. If arbitrage-enforced pricing does not work for a real options project, the Black-Scholes formula or risk-neutral valuation do not apply.

---

[1] If short selling of the real options is not possible, then arbitrageurs cannot earn profit by short the real options and long the replicating portfolio, though they can earn profit by long real options and short the replicating portfolio. This will make the price of the real options greater than or equal to the price for the replicating portfolio, rather than equal to the price of the replicating portfolio, and thus the arbitrage-enforced price does not hold.

4.6.2. What is the definition of real options?

People have different definitions of real options. To some extent, we do not even have a consensus of what are real options. Following is a partial list of different definitions:

- "In a narrow sense, the real options approach is the extension of financial option theory to options on real (nonfinancial) assets." (Amram and Kulatilaka, 1999)
- "Similar to options on financial securities, real options involve discretionary decisions or rights, with no obligations, to acquire or exchange an asset for a specified alternative price." (Trigeorgis, 1996)
- "Opportunities are options – right but not obligation to take some action in the future." (Dixit and Pindyck, 1995)
- "A real option is the right, but not the obligation, to take an action (e.g. deferring, expanding, contracting, or abandoning) at a predetermined cost called the exercise price, for a predetermined period of time – the life of the option." (Copeland and Antikarov, 2001)
- "In fact, it is possible to view almost any process that allows control as a process with a series of operational options. These operational options are often termed real options to emphasize that they involve real activities or real commodities, as opposed to purely financial commodities, as in the case, for instance, of stock options." (Luenberger, 1998)

Above definitions agree that options are rights not obligations. The key difference among the definitions lies in the scope of real options, from assets in a narrow sense to actions in a broad sense. If we insist that real options are application of financial options theory to nonfinanical assets, real options theory cannot be applied beyond the boundary where the "no arbitrage" assumption is valid. As designers of engineering systems, we think of real options in a broad sense that is close to Luenberger's definition – focusing on the trait of right not obligation and extending the real options concept in a more abstract way. And thus physical flexible design in an engineering systems can be thought of as real options, not only the engineering project as an investment opportunity as a whole.

Following the narrow and broad senses of definition of real options, there are two ways to understand the key contributions of real options concept:

- The nice theory of "no arbitrage" and risk-neutral valuation of assets that avoids the trouble to find out the correct risk-adjusted discount rate; or
- Defining the basic unit of flexibility analysis for any action or asset, that is, options (right not obligation)

We have proven the first contribution hardly stands for many real options "in" projects cases, if not most. Now let us examine closely the second argument of the contribution.

4.6.3. Options define flexibility

What is flexibility? How should it be measured? How should it be valued? Without a clearly defined basic unit of flexibility, it is hard to study it in an organized fashion.

Options concept neatly defines the basic unit of flexibility. The concept of real options is a right, but not obligation, to do something for a certain cost within or at a specific period of time. This concept models flexibility as an asymmetric right and obligation structure for a cost within a time frame. This is the basic structure of human decision making – take advantage of upside potential or opportunities and avoid downside risks. We can construct complex flexibility using the basic unit of real options.

Does Decision Analysis provide a means to structure flexibility? See the decision tree in Figure 4-7. The tree structure represents the flexibility to choose among Project 1, Project 2, Project 3 and Do Nothing. To a certain extent, a decision tree defines flexibility[1], but it has some inadequacy:

---

[1] Trigeorgis (1996) points out that decision tree analysis is "practically useful in dealing with uncertainty and with the modeling of interdependent variables and decisions, but they stumble on the problem of the appropriate discount rate."

- It aims at an expected value of the projects. This is over simplified with respect to the study of flexibility and human initiatives in risk management. It does not analyze each separate option and lose sight of the intricacy of flexibility.
- It could easily grow messy, and make analysts lose sight of the most important issues and choices.
- Decision tree discretize possibilities, but options analysis works with a continuous distribution and obtain more accurate and convincing results.

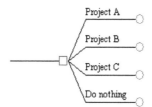

Figure 4-7 Decision Tree Analysis

Instead, a real option can serve as a basic unit to model flexibility. Real options can be stacked together to describe complex flexibility. For example, the decision tree in Figure 4-7 can be defined as a portfolio of three mutually exclusive call options on Project 1, 2, 3. Flexibility is a portfolio of real options.

Moreover, in comparison with decision tree analysis, real options analysis compares the value with and without options to get the value of options, helps people keep focus on the most important options, and values projects based on a continuous probability distribution of events.

Is real options "in" projects analysis is merely a fancy name for decision analysis? Doesn't it catch the essence of financial options theory that circumvents the problem of deciding appropriate discount rates? This argument looks plausible, but real options and decision analysis are different. Real options are building blocks to describe flexibility, and can be thought of as a formal way to define flexibility. Decision analysis is a way to organize different decision alternatives and possible outcomes to assist decision. Decision analysis is merely a tool and real options analysis is a way of thinking to understand, organize, summarize, and quantify flexibility.

In practice, real options theory has been extended into many areas where arbitrage-enforced pricing does not hold. The issue is not whether it is a **correct** real options valuation; there are some merits in such extension. Options definition has nothing to do with arbitrage-enforced pricing. It is broader. If they are financial options, we can use arbitrage-enforced pricing; if they cannot be valued by arbitrage-enforced pricing, they are still options, and they are still an interesting and useful way to define flexibility. This is the reason why real options grow more and more popular, despite the fact that some ingenious part of financial options theory is sometimes not valid in real options.

Chapter 5 Valuation of Real Options "in" Projects

Engineers increasingly recognize the great value of real options in addressing intrinsic uncertainties facing large-scale engineering systems and, more importantly, are learning to manage the uncertainties proactively. This book explores how to build real options analysis into the physical design of engineering systems, or real options "in" projects, which requires us to adapt financial real options theory and develop new tools.

Besides knowledge of technology, there are more difficulties facing the analysis of real options "in" projects: Financial options are well-defined contracts that are traded and that need to be valued individually. But real options "in" projects are fuzzy, complex, and interdependent: To what extent is there a predetermined exercise price? What is the time to expire? Moreover, it is not obvious the usefulness to value each element that provides flexibility. Real options "in" projects are likely to be path-dependent. For example, the capacity of a thermal power system at some future date may depend on the evolutionary path of electricity use. If the demands on the system have been high in preceding periods, the electric utility may have been forced to expand to meet that need, as it might not have done if the demand had been low. Real options for public services may thus differ fundamentally from stock options, whose current value only depends on the prices at that time. The evolutionary path of a stock price does not matter. Its option value is path-independent. This is not true for many real options.

Real options "in" projects are likely to be highly interdependent, compound options. Their interactions need to be studied carefully as they may have major consequences for important decisions about the design of the engineering system. The associated interdependency rapidly increases the complexity and size of the computational burden.

To develop a method for building real options "in" physical projects, the book offers suggestions for addressing the above difficulties:

It proposes to identify candidate real options "in" projects by screening and simulation models. This is important because, in an interdependent system, it may not be obvious where flexibility in the system may be most valuable. The book focuses on developing the most appropriate designs of flexibility and building up suitable contingency plans for dealing with future uncertainties.

To simplify the highly complicated path-dependent problem, it divides the decision time horizon into a small number of periods, then solves the path-dependent problem by a timing model using stochastic mixed-integer programming. This process also deals with the compound options difficulty.

5.1. Analysis framework

The analysis of real options "in" projects is inspired by the standard procedures for water resources planning described by Major and Lenton (1979). The standard procedures embody a series of models to generate satisfactory solution for the plan. Because of the size of the problem, both in terms of number of parameters and number of uncertain variables, a single model giving the optimal solution is too complex to establish. So people divide the modeling into a series of models and get a satisfactory solution rather than search the best plan among all possible plans. As Herbert Simon pointed out: because of the astronomical amount of extrinsic information and human's limited intrinsic information process capacity, in real decision making, people do not search for an optimal decision, instead, people stop looking for better decisions once reaching a satisfactory decision. The process - that divides the decision process into several consecutive models and search for a satisfactory solution rather than an optimal solution - is in accordance with the nature of human decision-making in a very complex and uncertain environment. Specifically, the standard water resources planning procedures divide the process into: a deterministic screening model that identifies the possible elements of the system that seem most desirable; a simulation model that explores the performance of candidate designs under stochastic loads; and a timing model that defines an optimal sequence of projects.

The process of analysis for real options "in" projects modifies these traditional elements. At a higher level, it divides the analysis into 2 phases as indicated in Figure 5-1: options identification, and options analysis.

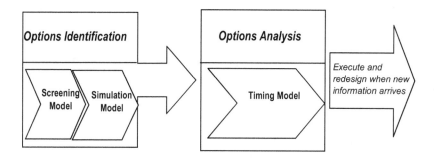

Figure 5-1: Process for Analysis of Real Options "in" Projects

5.2. Options identification

For real options "in" projects, the first task is to define the options. This is in contrast with financial options, whose terms (exercise price, expiration day, and type such European, American or Asian) are clearly defined. For financial options, the main task is to value the option and develop a plan for its exercise. For real options "in" projects, it is only possible to analyze the options to show their value and develop a contingency plan for the management of the projects, after the options have been identified. This first task for real options "in" projects is not trivial.

5.2.1. Screening model

The options "in" projects for an engineering system are complex. It is not obvious how to decide their exercise price, expiration day, current price, or even to identify the options themselves. An engineering system involves a great many choices about the date to build, capacity, and location, etc. The question is: which options are most important and justify the resources needed for further study?

The screening model is established to screen out the most important variables and interesting real options (flexibility). The screening model is a simplified, conceptual, low-fidelity model for the system. Without losing the most important issues, it can be easily run many times to explore an issue, while the full, complete, high-fidelity model is hard to establish and costly to run many times. From another perspective on the screening model, we can think of it as the first step of a process to reduce the design space of the system. The design space is extremely big and the possibility for future realization of exogenous uncertain factors is also extremely big. Therefore, we cut the design space smaller and smaller in steps, rather than using a holistic model to accomplish all the results in one run. The screening model is the first cut that focuses on the important issues and is low fidelity in nature, like looking at the system at the 30,000 feet height for an overview. The screening model may be simplified in a number of ways. If an aspect simplified is important in nature, we should design follow-on models to the screening model to study that aspect in depth. Note when feedback exists in the system, the screening model has to carefully take care of the feedback; otherwise, it may produce misleading or erroneous conclusions.

For example, in our case example of water resources planning (details see Chapter 6), the screening model is a non-linear programming model that optimizes the system assuming steady state, i.e. all projects are built all at once. We also simplify the problem by regarding it as a deterministic problem, i.e., taking out the stochasticity of the water flow and electricity price. The screening model does not consider all the complexities of the system; it considers a large numbers of possibilities, screens out most of them, and focuses attention on the promising designs. With such simplifications, we are able to focus our attention on identifying the most interesting flexibility, and leave the scrutiny of the aspects in the following models and studies.

Specifically, a screening model can be a linear (or nonlinear) programming model:

$$\text{Max:} \qquad \sum_{j}(\beta_j Y_j - c_j Y_j) \qquad\qquad \text{Equation 5-1}$$

$$\text{s.t.} \qquad \mathbf{TY} \geq \mathbf{t} \qquad\qquad\qquad \text{Equation 5-2}$$

$$\mathbf{EY} \geq \mathbf{e} \qquad\qquad\qquad \text{Equation 5-3}$$

Y_j are the design parameters. The objective function (Equation 5-1) calculates the net benefit, or the difference between the benefits and costs, where β_j and c_j are the benefit and cost coefficients. Usually we measure benefits in money terms, though sometimes we do so in other measures, e.g. species saved, people employed, etc. Constraints (Equation 5-2) and (Equation 5-3) represent technical and economic limits on the engineering systems, respectively.

Any parameter in the formulation could be uncertain. There are economic uncertainties in $\mathbf{E}$, $\mathbf{e}$, β_j, or c_j and technical uncertainties in $\mathbf{T}$ or $\mathbf{t}$. After identifying the key economic uncertainties, we can use them as underlying to build up real options analysis as illustrated in the case example in Chapter 6.

To identify the elements of the system that seem most promising for options, we execute a form of sensitivity analysis as follows: run the screening model using a range of values for key underlying uncertain parameters, such as the price of electricity; compare the resulting sets of projects that constitute optimal designs for each set of parameters used;

and the design elements that vary across the sets are these that may or may not be good real options; conversely, the design elements that are included for all sets, that are insensitive to uncertainty, or design elements where settings are always constant do not present interesting real options.

The resulted options have two sources of flexibility value:
- Value of timing. Some part of the project may be deferred. These represent timing options. Its implementation depends on the realized uncertain variables. Since it can catch upside of the uncertainties by implementation, and avoid downside of the uncertainties by holding implementation. Such timing options have significant value by themselves.
- Value of flexible design. Some part of the project may present distinct designs given various realization of uncertainties, compared to the timing options whose design are the same whenever they are built.

An option can contain one or both sources of flexibility value.

5.2.2. Simulation model

The simulation model tests several candidate designs from runs of the screening model. It is a high fidelity model. Its main purpose is to examine, under technical and economic uncertainties, the robustness and reliability of the designs, as well as their expected benefits. Such extensive testing is hard to do using the screening model. After using the simulation model, we find a most satisfactory configuration with design parameters $(\overline{Y_1}, \overline{Y_2}, ..., \overline{Y_j})$ in preparation for the options analysis.

In standard water resources planning, the simulation model involves many years of simulated stochastic variation of the water flows, generated on the basis of historical records. This process leads to a refinement of the designs identified by the screening model. For the analysis of real options "in" water resources systems, we propose to modify this standard simulation process. Specifically, we will simulate the combined effect of stochastic variation of hydrologic and economic uncertain parameters.

If the time series of the water flow consisted of the seasonal means repeating themselves year after year (no shortages with regard to the design obtained by the screening model) and the price of electricity were not changing, the simulation model should provide the same results as the screening model. But the natural variability of water flow and electricity price will make the result (net benefit) of each run different, and the average net benefit is not going to be the same as the result from the screening model. The simulated results should be lower because the designs are not going to benefit from excess water when water is more than the reservoir can store. Thus occasional high levels of water do not provide compensation for lost revenues by occasional low levels of water. Due to these uncertainties, the economies of scale seemingly apparent under deterministic schemes are reduced.

5.3. Options analysis

After identifying the most promising real options "in" projects, designers need a model that enables them to value the set of options and develop a contingency strategy for their exercise. In contrast to standard financial options analysis, more characteristics are required for the analysis of real options "in" projects, such as technical details and interdependency/path-dependency among options.

This book proposes a model based on the scenarios established by a binomial lattice. In essence, it proposes a new way to look at the binomial tree, recasting it in the form of a stochastic mixed-integer programming model. The idea is to:

Maximize: binomial tree

Subject to: constraints consisting of 0-1 integer variables representing the exercise of the options (= 0 if not exercised, =1 if exercised)

5.3.1. Using integer programming to solve a binomial tree

By simple examples on financial options, we would illustrate the basic idea of using stochastic mixed-integer programming model to value options.

Important variables for options valuation are as follows

S: Stock Price

K: Exercise Price

T: Time to Expiration

r: Risk free interest rate

σ: Volatility

ΔT: Time interval between nodes

Important Formulas for Binomial Tree Model include:

$$u = e^{\sigma\sqrt{\Delta T}}$$
$$d = e^{-\sigma\sqrt{\Delta T}}$$
$$p = \frac{e^{r\Delta T} - d}{u - d}$$

At each node of a binomial tree, the calculation is as Table 5-1. Note this is for the valuation of American options, and p is the risk-neutral probability, the same as p_u in Section 3.3.3.

Table 5-1: Decision on each node of a binomial lattice

Stock Price	S
Exercise Value	$S - K$ (for call); $K - S$ (for put)
Hold Value	$\dfrac{Option\_Value\_if\_price\_up \cdot p_u + Option\_Value\_if\_price\_down \cdot p_d}{e^{r \cdot \Delta T}}$
Options Value	*Max (Exercise Value, Hold Value)* (0 for the last period since no options value at expiration)

Now we will compare an ordinary binomial tree and an integer programming binomial tree. The interesting part is to compare the option value from the binomial tree and the optimal value from the integer programming, as well as the "exercise or not" result for each node of the binomial tree and the value of 0-1 integer variables in the optimal solution of the integer program. The American option is of special interest because we want to examine if the integer program can correctly identify the case of early exercise before the last period.

For example, the parameters for an American call option are S = $20, K = $21, T = 3 years, r = 5% per year, σ = 30%, ΔT = 1 year. The binomial tree is as Table 5-2, and the value of the option is $5.19.

Now let us use Integer programming to value this binomial tree. The node *im* on a binomial tree is indexed in the following way: *i* represents the *ith* stage, *m* represents the *mth* node for a specific stage. Because of the nice feature of recombination of a binomial tree when there is path independence, the number of nodes at ith time point is exactly *i*, so *m* takes the number from 1 to *i*. Please refer to Figure 5-2.

At node im, let S_{im} denote the stock price, E_{im} denote the exercise value, H_{im} denote the hold value, V_{im} denote the option value, R_{im} be a 0-1 integer variable denoting whether the option is exercised at node *im*, where 0 is not exercise and 1 is exercise. The number of stages is *n*.

Table 5-2: Binomial tree for the American call option

	Period 1	Period 2	Period 3	Period 4
Stock Price	20.00	27.00	36.44	49.19
Exercise Value	-1.00	6.00	15.44	28.19
Hold Value	5.19	9.34	16.47	0.00
Option Value	5.19	9.34	16.47	28.19
Exercise or not?	No	No	No	Yes
Stock Price		14.82	20.00	27.00
Exercise Value		-6.18	-1.00	6.00
Hold Value		1.41	2.91	0.00
Option Value		1.41	2.91	6.00
Exercise or not?		No	No	Yes
Stock Price			10.98	14.82
Exercise Value			-10.02	-6.18
Hold Value			0.00	0.00
Option Value			0.00	0.00
Exercise or not?			No	No
Stock Price				8.13
Exercise Value				-12.87
Hold Value				0.00
Option Value				0.00
Exercise or not?				No

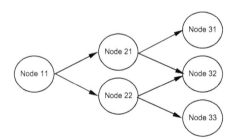

Figure 5-2: Node representation for a binomial tree

The objective function is to get the maximum value of V_{11} at the beginning node. The option value V_{im} is specified by

$$V_{im} = E_{im} \cdot R_{im} + H_{im} \cdot (1 - R_{im})$$

Since the programming maximizes the value, its final result will satisfy that V_{im} is the maximum of E_{im} and H_{im}.

The exercise value E_{im} for a call options is

$$E_{im} = S_{im} - K$$

The hold value for the last time point is 0, or $H_{n,m} = 0$. For $i < n$, the hold value

$$H_{im} = \frac{V_{i+1,m} \cdot p + V_{i+1,m+1} \cdot (1-p)}{e^{r \cdot \Delta T}}$$

We are using continuous compounding here.

The stock price S_{im} at node im is defined by the following formula

$$S_{im} = S_{11} e^{(i+1-2m)\sigma\sqrt{\Delta T}}$$

where S_{11} is the current stock price.

Complete formulation of the integer programming problem is as follows:

Maximize $\quad V_{11}$

Subject to $\quad V_{im} = E_{im} \cdot R_{im} + H_{im} \cdot (1 - R_{im}) \qquad \forall i; m = 1,...,i$

$\qquad\qquad E_{im} = S_{im} - K \qquad\qquad\qquad\qquad\quad \forall i; m = 1,...,i$

$\qquad\qquad H_{im} = \dfrac{V_{i+1,m} \cdot p + V_{i+1,m+1} \cdot (1-p)}{e^{r \cdot \Delta T}} \quad i = 1,...,n-1; m = 1,...,i$

$\qquad\qquad H_{n,m} = 0 \qquad\qquad\qquad\qquad\qquad\quad \forall m$

$\qquad\qquad S_{im} = S_{11} e^{(i+1-2m)\sigma\sqrt{\Delta T}} \qquad\qquad\quad \forall i; m = 1,...,i$

Solve the integer program using GAMS or any other Mixed-Integer Programming solver (see GAMS source code in Appendix 5A), the maximum value of the objective function is 5.19. The values for 0-1 integer variables are as Table 5-3: 1 means exercise. The result of option value and contingent exercise decisions correspond exactly to the ordinary binomial tree as Table 5-2.

Table 5-3: Result of the stochastic programming for the American call option

Rij	i = 1	i = 2	i = 3	i = 4
j = 1	0	0	0	1
j = 2		0	0	1
j = 3			0	0
j = 4				0

Such formulation seems unnecessarily complicated for a simple financial option. But for complex and highly interdependent real options "in" projects, we can specify the relationship of options using the 0-1 integer variable constraints. Without integer programming, a binomial tree for a path-dependent real option "in" projects may be too messy to build. With technical, budget, and real options constraints, a stochastic mixed-integer programming model accounts for highly complex and interdependent issues, and delivers both a valuation of the options and a contingency strategy. Details see the coming sections.

5.3.2. Stochastic mixed-integer programming and real options constraints

This section develops a general formulation for the analysis of real options "in" projects, especially those with path-dependency.

The stochastic mixed-integer programming assumes that the economic uncertain parameters in $\mathbf{E}$, $\mathbf{e}$, β_j, or c_j in objective function Equation 5-1 and constraints Equation 5-2 and Equation 5-3 evolve as discrete time stochastic processes with a finite probability space. A scenario tree is used to represent the evolution of an uncertain parameter (Ahmed, King, and Parija, 2003). Figure 5-3 illustrates the notation. The nodes k^1 in all time stages i constitute the states of the world. δ_i denotes the set of nodes corresponding to time stage i. The path from the root node 0 at the first stage to a node k is denoted by $P(k)$. Any node k in the last stage n is a terminal node. The path $P(k)$ to a terminal node represents a scenario, a realization over all periods from 1 to the last stage n. The number of terminal nodes Q corresponds to all Q scenarios. Note there is no recombination structure in this tree representation (each node except the root has a unique parent node). For example, we will break a binomial tree structure as in Figure 5-4, where S is the value of the underlying asset, u is the up factor, and d is the down factor.

A joint realization of the problem parameters corresponding to scenario q is denoted by

$$\omega^q = \begin{pmatrix} \omega_1^q \\ ... \\ \omega_T^q \end{pmatrix},$$

---

[1] Note that here we use a single dimension to represent a node, rather than two dimensions in Figure 5-2 where nodes are represented by 11, 21, 22, 31, 32, 33.

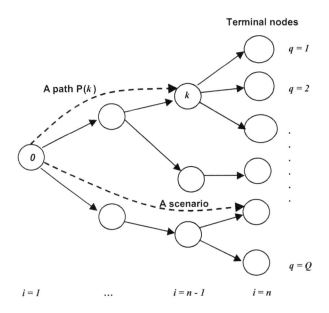

Figure 5-3: Scenario tree

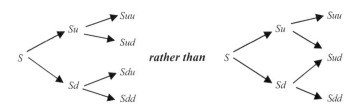

Figure 5-4: Breaking path-independence of a binomial tree

where ω_i^q is the vector consisting of all the uncertain parameters for time stage i in scenario q. p^q denotes the probability for a scenario q. The real options decision variables corresponding to scenario q is denoted by

$$\mathbf{R^q} = \begin{pmatrix} R_1^q \\ ... \\ R_T^q \end{pmatrix},$$

where R_i^q is the decision on the option at time stage i in scenario q. 0 denotes no exercise and 1 denotes exercise.

At any intermediate stage i, the decision maker cannot distinguish between any scenario passing through the same node and proceeding on to different terminal node, because the state can only be distinguished by information available up to that time stage. Consequently, the feasible solution R_i^q must satisfy:

$$R_i^{q_1} = R_i^{q_2} \qquad \forall (q_1, q_2) \text{ through node } k, \forall k \in \delta_i, \forall i = 1,...,n$$

where q_1 and q_2 represent two different scenarios that share the same path through node k. These constraints are known as non-anticipativity constraints.

To illustrate the use of the above approach, we apply it to some standard financial options. The formulation is:

Max $\quad \sum_q p^q \cdot (\sum_{i=1}^{n} E_i^q \cdot R_i^q \cdot e^{-r \cdot \Delta T \cdot (i-1)})$ Equation 5-4

s.t. $\quad E_i^q = S_i^q - K$ (American call) $\quad \forall i, q$ Equation 5-5

$\qquad \sum_i R_i^q \leq 1 \qquad\qquad \forall q$ Equation 5-6

$\qquad R_i^{q_1} = R_i^{q_2} \qquad\qquad \forall (q_1, q_2) \text{ through node } k, \forall k \in \delta_i, \forall i = 1,...,n$ Equation 5-7

$\qquad R_i^q \in \{0,1\} \qquad\qquad \forall i, q$ Equation 5-8 where S_i^q is the value of the underlying asset at time stage i in scenario q, K is the exercise price, r is the risk-free interest rate, and ΔT is the time interval between two consecutive stages.

The objective function (Equation 5-4) is the expected value of the option across all scenarios. Constraint (Equation 5-5) can be any equation that specifies the exercise condition. Constraint (Equation 5-6) makes sure that any option can only be exercised at most once in any scenario. Constraint (Equation 5-7) is the non-anticipativity constraints. We call constraints (Equation 5-6) and (Equation 5-7) real options constraints. Equation 5-8 enforces a binary decision on exercise, e.g. $R_i^q = 0.5$ (partial exercise) in not allowed.

To illustrate and validate the above formulation, consider an example of an American put option without dividend payment. For this case, unlike similar call options, it may be optimal to exercise before the last period. The variables for this example are S = $20, K = $18, r = 5% per year, σ = 30%, ΔT = 1 year, and time to maturity T = 3 years. Up factor u = 1.35, down factor d = 0.74. A standard binomial lattice gives the value of the put option as $2.20 as in Table 5-5.

Now considering the reformulated problem according to equations (Equation 5-4) to (Equation 5-8). Solve it using GAMS© (GAMS source code can be found in Appendix 5B. Note the difference between formulations in Appendix 5A and 5B. They are two different formulations to value options using stochastic mixed-integer programming), the maximum value of the objective function is also 2.20. The optimal solution of 0-1 variables is shown in Table 5-4. Since 1 means exercise, the result exactly corresponds to that of the ordinary binomial tree (Table 5-5). Note there is an exercise in scenarios 7 and 8 that is not at the last time point. This means the formulation can successfully find out early exercise points and define contingency plans for decision makers.

Table 5-4: Stochastic programming result for the example American put

Scenario	Stock Price Realization					Decision			
	$i = 1$	$i = 2$	$i = 3$	$i = 4$	Probability	i = 1	i = 2	i = 3	i = 4
$q = 1$	S 20.00	Su 27.00	Suu 36.44	Suuu 49.19	0.132	0	0	0	0
$q = 2$	S 20.00	Su 27.00	Suu 36.44	Suud 27.00	0.127	0	0	0	0
$q = 3$	S 20.00	Su 27.00	Sud 20.00	Sudu 27.00	0.127	0	0	0	0
$q = 4$	S 20.00	Su 27.00	Sud 20.00	Sudd 14.82	0.123	0	0	0	1
$q = 5$	S 20.00	Sd 14.82	Sdu 20.00	Sduu 27.00	0.127	0	0	0	0
$q = 6$	S 20.00	Sd 14.82	Sdu 20.00	Sdud 14.82	0.123	0	0	0	1
$q = 7$	S 20.00	Sd 14.82	Sdd 10.98	Sddu 14.82	0.123	0	0	1	0
$q = 8$	S 20.00	Ds 14.82	Sdd 10.98	Sddd 8.13	0.118	0	0	1	0

Table 5-5: Binomial tree for the example American put

	Period 1	Period 2	Period 3	Period 4
Stock Price	20.00	27.00	36.44	49.19
Exercise Value	-2.00	-9.00	-18.44	-31.19
Hold Value	2.20	0.69	0.00	0.00
Option Value	2.20	0.69	0.00	0.00
Exercise or not?	No	No	No	No
Stock Price		14.82	20.00	27.00
Exercise Value		3.18	-2.00	-9.00
Hold Value		4.00	1.48	0.00
Option Value		4.00	1.48	0.00
Exercise or not?		No	No	No
Stock Price			10.98	14.82
Exercise Value			7.02	3.18
Hold Value			6.15	0.00
Option Value			7.02	3.18
Exercise or not?			Yes	Yes
Stock Price				8.13
Exercise Value				9.87
Hold Value				0.00
Option Value				9.87
Exercise or not?				Yes

5.3.3. Formulation for the real options timing model

The stochastic mixed-integer programming reformulation is much more complicated than a simple binomial lattice. But such a reformulations empowers analysis of complex path-dependent real options "in" projects for engineering systems.

Technical constraints in the screening model are modified in the real options timing model. Since the screening and simulation models have identified the configuration of design parameters, these are no longer treated as decision variables. On the other hand, the timing model relaxes the assumption of the screening model that the projects are built together all at once. It decides the possible sequence of the construction of each project in the most satisfactory designs for the actual evolution of the uncertain future.

$\overline{Y}$ is the most satisfactory configuration of design parameters obtained by the "options identification" stage, it is a vector $(\overline{Y_1}, \overline{Y_2}, ..., \overline{Y_j})$ corresponding to j design parameters. The real options decision variable corresponding to scenario q is expanded to a matrix:

$$\mathbf{R}^q = \begin{bmatrix} R_{11}^q & \cdots & R_{1j}^q \\ \vdots & & \vdots \\ R_{i1}^q & \cdots & R_{ij}^q \end{bmatrix}, \quad R_{ij}^q \in \{0,1\}$$

R_{ij}^q denotes the decision on whether to build the feature according to jth design parameter for ith time stage in scenario q. The objective function Equation 5-4 corresponding to scenario q is denoted by $f^q(\cdot)$. p^q and $f^q(\cdot)$ are derived from the specific scenario tree based on the appropriate stochastic process for the subject under study. The real options constraints Equation 5-6 to Equation 5-7 are concisely denoted by φ. Most importantly, the objective function is modified to get the expected value along all scenarios.

The real options timing model formulation is as follows:

$$\text{Max} \quad \sum_{q} p^{q} \cdot f^{q}(\mathbf{R}^{q}, \overline{\mathbf{Y}})$$

$$\text{s.t.} \quad \mathbf{T} \begin{bmatrix} \overline{Y}_1 R_{i1}^{q} \\ : \\ \overline{Y}_j R_{ij}^{q} \end{bmatrix} \geq \mathbf{t} \quad \text{and} \quad \mathbf{E} \begin{bmatrix} \overline{Y}_1 R_{i1}^{q} \\ : \\ \overline{Y}_j R_{ij}^{q} \end{bmatrix} \geq \mathbf{e} \quad \forall q, i$$

$$\mathbf{R}^{q} \in \varphi$$

$$R_{ij}^{q} \in \{0,1\} \qquad \qquad \forall q, i, j$$

In short, the formulation has an objective function averaged over all the scenarios, subject to three kinds of constraints: technical, economic, and real options. By specifying the interdependencies by constraints, we can take into account highly complex relationship among projects.

Some integer programming model running considerations

A key consideration in solving a stochastic mixed-integer programming is whether a result is a global or local optimum. Because it is a integer programming problem, it is not often a simple task to prove that the result is a global optimum.

One way to prove a global optimum, when we are extremely lucky, is to use the relaxed mixed integer program to solve the problem, if the decision variables X_{ij} in the optimal design take value exactly either 0 or 1, we can confidently say that the optimal value of the objective function is the global optimum (for more discussion, see Bertsimas and Tsitsiklis, 1997). Please note that all the constraints in the formulations are linear. So the local optimum found is also the global optimum. Because the decision variable X_{ij} in the relaxed form takes 0 or 1, it satisfies the requirements of mixed integer programming.

In practice, however, it is rarely straightforward to prove that the result is a global optimum. This is one of the reasons why people would not use this way to value financial options (but this formulation provides significant advantages for real options "in" projects where special path-dependency presents). Of course, system designers should run many times of optimization with different initial conditions when they cannot be sure of the global optimum. By doing do, designers can endeavor to make the result as close as possible to the global solution, given resources limitations.

It may be hard to find a general solution for the real options timing model because of the special structure of the technical and economic constraints. Nevertheless, integer programming improves solutions to highly complex and interdependent real options that cannot be solved by ordinary binomial trees. Ordinary binomial trees depend on simple human observations. Here simple human observations mean, when there is no dependency between nodes, people can do simple math to optimize on the node. When there is no dependency among nodes, it is possible to optimize on each node and roll back to get the option value. When dependency exists, this simple approach no longer works. A stochastic mixed-integer programming at least provides a local optimum better than the results from conventional approaches or human intuition.

When the optimization returns an infeasible solution, we should try with different initial values of X_{ij}, especially try some X_{ij} at the last period of time to be 1's. Besides, when it is hard to find a feasible solution, we can use relaxed form of the problem to see what happens.

5.4. A case example on satellite communications

system

This is a case example to show how the models developed can deal with the valuation/design of flexibility in large-scale engineering systems, especially in path-dependent/interdependent cases. The case is drawn from the paper of "Staged Deployment of Communications Satellite Constellations in Low Earth Orbit" by de Weck, de Neufville, and Chaize (2004).

The case discusses Low Earth Orbit (LEO) constellations of communications satellites. The size of the market is close to what Iridium system originally expected, about 3 million subscribers. The most important architectural design decisions for the system are captured by a design vector $Y = [h, \varepsilon, P_t, D_A, ISL]$ [1], where h is the orbital altitude in Kilometers, ε is the minimum elevation angle in Degrees, P_t is the satellite transmit power in Watts, D_A is the antenna diameter in meters, and ISL is the use of inter satellite links, a binary variable, 0 or 1.

In a design space that relates life cycle cost and capacity of architectures, we can find the Pareto frontier that represents non-dominated architectures. With the target capacity, we can find the corresponding architecture on the Pareto frontier. This is the traditional way to decide the best architecture. The best traditional architecture for this case $\overline{Y}^{trad} = [800, 5, 600, 2.5, 1]$, or 800 Kilometers orbital altitude, 5 degrees minimum elevation angle, 600 Watts satellite transmit power, 2.5 meters antenna diameter, and with the use of inter satellite links. It has a discounted life circle cost of LCC = $2.01 Billion. This architecture can serve up to 2.82 million subscribers. Note this constellation is quite similar to Iridium, $\overline{Y}_{Iridium} = [780, 8.2, 400, 1.5, 1]$ with a market capacity of 3 million subscribers.

---

[1] All the notations here follow the original paper by de Weck, de Neufville, and Chaize (2004).

5.4.1. Market demand uncertainty

The constellation is assumed to have the same life time designed as Iridium, 10 years. The initial demand $D_{Initial}$ is set to be 0.5 million subscribers. A geometric Brownian motion is assumed for the demand growth. The drift rate μ is taken to be 20% per year, and the volatility σ is taken to be 70%. Because of the path-dependent nature of the problem, the binomial tree recombination structure is broken as in Figure 5-4. We consider the same number of stages as in the original paper by de Weck, de Neufville, and Chaize (2004), or 5 time stages, so the time step ΔT is set to be 2.5 years.

Up and down factors are respectively:

$$u = e^{\sigma\sqrt{\Delta T}} = 3.02$$
$$d = e^{-\sigma\sqrt{\Delta T}} = 0.33$$

The probabilities for moving up and down are respectively:

$$p_u = \frac{e^{\mu\Delta T} - d}{u - d} = 0.489$$
$$p_d = 1 - p_u = 0.511$$

We use D_i^q (in million subscribers) to denote the demand for stage i under scenario q. p^q is the probability for scenario q to happen. The different scenarios are as Table 5-6.

Table 5-6 Evolution of demand for satellite system

	Demand (in million subscribers)					Probability (p^q)
	Year 0 $(i = 1)$	Year 2.5 $(i = 2)$	Year 5 $(i = 3)$	Year 7.5 $(i = 4)$	Year 10 $(i = 5)$	
$q = 1$	0.50	1.51	4.57	13.84	41.85	0.057
$q = 2$	0.50	1.51	4.57	13.84	4.57	0.060
$q = 3$	0.50	1.51	4.57	1.51	4.57	0.060
$q = 4$	0.50	1.51	4.57	1.51	0.50	0.062
$q = 5$	0.50	1.51	0.50	1.51	4.57	0.060
$q = 6$	0.50	1.51	0.50	1.51	0.50	0.062
$q = 7$	0.50	1.51	0.50	0.17	0.50	0.062
$q = 8$	0.50	1.51	0.50	0.17	0.05	0.065
$q = 9$	0.50	0.17	0.50	1.51	4.57	0.060
$q = 10$	0.50	0.17	0.50	1.51	0.50	0.062
$q = 11$	0.50	0.17	0.50	0.17	0.50	0.062
$q = 12$	0.50	0.17	0.50	0.17	0.05	0.065
$q = 13$	0.50	0.17	0.05	0.17	0.50	0.062
$q = 14$	0.50	0.17	0.05	0.17	0.05	0.065
$q = 15$	0.50	0.17	0.05	0.02	0.05	0.065
$q = 16$	0.50	0.17	0.05	0.02	0.01	0.068

5.4.2. Formulation of the problem

R_{is}^q is a 0-1 variable indicating whether or not architecture s is built for time period i under scenario q. In this satellite technology, changes of capacity imply change of architectures. In this case, we consider $i = 5$, $q = 16$, and $s = 5$. With 5 stages, the problem is not trial so that it cannot be calculated back-of-the-envelope.

The problem can be formulated as

$$\text{Min} \quad \sum_q p^q \sum_i (\sum_s C_s R_{is}^q (1+r)^{(\Delta T - i \cdot \Delta T)} (1 + a - a \cdot (\sum_s R_{is}^q)^b))$$

$$\text{S.t.} \quad \sum_s (Cap_s \cdot \sum_{j=1}^{i} R_{js}^q) \geq D_i^q \qquad \text{Equation 5-9}$$

$$\sum_{j=1}^{i} R_{j,s-1}^{q} \leq R_{js}^{q} \qquad \forall i; \forall s = 2,3,4,5 \qquad \text{Equation 5-10}$$

$$\sum_{i} R_{is}^{q} \leq 1 \qquad \forall s,q \qquad \text{Equation 5-11}$$

$$R_{is}^{q_1} = R_{is}^{q_2} \qquad \forall (q_1,q_2) \text{ through node } k, \forall k \in \delta_i, \forall i = 1 \qquad \text{Equation 5-12}$$

$$R_{ij}^{q} \in \{0,1\} \qquad \forall q,i,j$$

The objective function describes the expected cost for the flexible strategy to meet the demand under various scenarios. We want to minimize the expected cost given that we can meet the demand. C_s is the cost coefficient for architecture s. The term $(1 + a - a \cdot (\sum_{s} R_{is}^{q})^{b})$ describes economies of scale. If two or more architectures are deployed together, there will be cost saving over the sum of the costs for the two or more separate architectures.

The constraint Equation 5-9 makes sure that the capacity always meets the demand. Cap_s represents the incremental capacity an evolutionary architecture design s adds over the previous design. The demand in some scenario is greater than the maximum total capacity considered (7.8 million subscribers, see Section 5.4.3.), e.g. when q = 1 and i = 5 in Table 5-6. To make the problem feasible to solve, we will have to make the demand D_i^q to be the maximum of the capacity if it is bigger than the maximum possible capacity and thus make the problem infeasible.

The constraint Equation 5-10 makes sure that a bigger architectural design can only build on smaller architectural designs. This constraint forbids skipping of stages, such as A1 to A3 directly. It allows two more consecutive architectures to be implemented together at the same time, for example A1 and A2 together.

The constraints Equation 5-11 and Equation 5-12 are the real options constraints we have developed.

5.4.3. Parameters

We consider three architecture designs as Table 5-7:

Table 5-7 Candidate architecture design for satellite system

Design ID	Architecture	Incremental Capacity (Subscribers)	Incremental Cost (B $)
A1	28 satellites, 4 planes, 1600 km altitude, 5 degrees	0.4 Million	0.25
A2	32 satellites, 4 planes, 1200 km altitude, 5 degrees	0.1 Million	0.15 over Design A1
A3	84 satellites, 7 planes, 1200 km altitude, 20 degrees	0.8 Million	0.7 over Design A2
A4	144 satellites, 9 planes, 800 km altitude, 20 degrees	1.4 Million	0.8 over Design A3
A5	364 satellites, 14 planes, 800 km altitude, 35 degrees	5.1 Million	4.9 over Design A4

As an approximation, the economies of scale factor a and b are taken to be 0.03 and 1.5, respectively.

5.4.4. The results

Using GAMS (the code see Appendix 5C), the resulting optimum cost is $1.90 Billion, $0.11 Billion less than the best traditional architecture. It is not the most important improvement of the staged deployment that it has a smaller expected cost than that of the best traditional architecture. The best improvement is it can take advantage of upside potential, while cut downside risks. The plan can serve up 7.8 million subscribers readily, compared to the best traditional architecture that can only serve up to 2.82 million subscribers. Meanwhile, we would first invest $0.25 Billion to test the market and, if the market is not rosy, we could only lose $0.25 Billion rather than $2.01 Billion as of investment for the best traditional architecture. The downside can be significantly cut. In comparison to what happened to Iridium and Globalstar, this way of analysis deserves serious attention.

With the GAMS results, we develop a contingency plan for the development of the satellite system as Figure 5-5. Note that architectures A1 and A3 are not in the plan. This is because of the considerations of economies of scale. Building a bigger system, the benefit of economies of scale sometimes overweighs the benefit of postponement of construction (because of the time value of money, investment later has a smaller present value).

Computational efficiency

A natural question is: how efficient is the mixed-integer programming method? One possible way to illustrate its computational efficiency is to compare it with the "brute force" method, the most readily available alternative algorithm to the mixed-integer programming method. The "brute force" model lists all possibilities and calculates values of each. The "brute force" approach would
    1) list all combinations
    2) check feasibility of each, discarding the infeasible ones
    3) evaluate remaining set.
Note we are not deciding in advance which combinations are "infeasible".

We estimate the computational implications of such an approach by estimating the number of combinations. Suppose the number of decision stages is i, the number of potential design choices is $(s + 1)$, [1] and there are 2 possible scenarios following each decision based on the binomial layout of scenarios. For the satellite communication system case example, there are 5 stages and 6 potential choices. So the possible number of combinations for the satellite communication case example is:

---

[1] The number of candidate designs is s, but there is also a choice to do nothing.

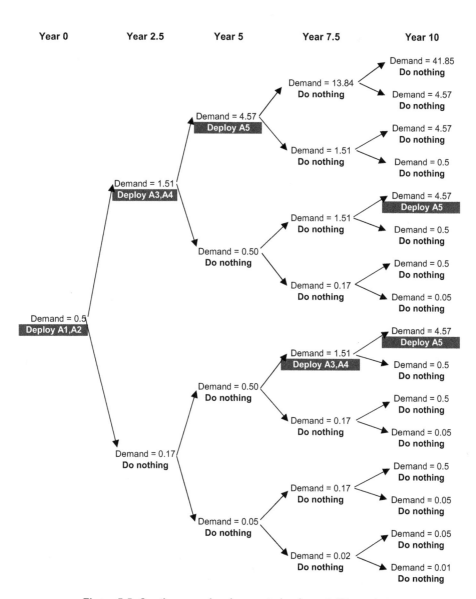

Figure 5-5: Contingency development plan for satellite system

$$6^5 \times 2^5 = 248832$$

We can put all 248832 combinations into the "brute force" model, then check feasibility of each given the path-dependency constraints (such as order of deployments) and other technical and market constraints, and finally evaluate all feasible combinations by calculating expected costs and taking the one with the lowest expected cost.

Another possible way to enumerate is to consider the order of deployments in the first place, and thus we can cut the number of combinations fed into the feasibility check. For example, Architecture A2 cannot be built before A1; and once A1 is built, we cannot build it again. So the number of possible decision choices varies on each path and node, though the number is less than or equal to 6. Given the fact that, in this case, architectures can be built together and nothing being built can be a choice, such enumeration is very messy and it is easy to make mistakes.

Compared the mixed-integer programming of the problem that is solved on my laptop (CPU Pentium III 650 MHz, RAM 384M) in less than 2 minutes. It is a bigger effort to list all the possible combinations, check the feasibility of each combination, and evaluate the expected cost for each combination and find the one with the minimum expected cost

5.4.5. Discussion

The above analysis is crude yet renders important results, especially the possibility to avoid big loss compared to the best traditional architecture - we can abandon future expansion plans if things turn sour. In real decision making, we can refine the model with smaller time intervals between stages, ΔT, and refine the scenario tree for the uncertain variables (though the basic formulation is the same). The Geometric Brownian motion may not the best stochastic model for the demand growth, and we may build alternative scenario trees based on better stochastic models. Moreover, the supply of services may impact the demand, and we can add feedback loop onto scenario trees to take into account the impact of supply on demand.

The model presented does not dig into the technology itself deeply. If we want to include in-depth technical considerations, the above model is readily expandable to deal with technical details. We can change the cost coefficient C_s to a function $C(h, \varepsilon, P_t, D_A, ISL, ...)$ of the important architectural variables, and add a group of technical constraints:

$$g(h, \varepsilon, P_t, D_A, ISL, ...) \leq 0$$

So the general formulation with more technical details for the satellite communications system problem is:

Min $\quad \sum_q p^q \sum_i (\sum_s C(h, \varepsilon, P_t, D_A, ISL, ...) R_{is}^q (1+r)^{(\Delta T - i \cdot \Delta T)} (1 - a \cdot (\sum_s R_{is}^q)^b))$

S.t. $\quad g(h, \varepsilon, P_t, D_A, ISL, ...) \leq 0$

$$\sum_s (Cap_s \cdot \sum_{j=1}^{i} R_{js}^q) \geq D_i^q$$

$$\sum_{j=1}^{i} R_{j,s-1}^q \leq R_{js}^q \qquad \forall i; \forall s = 2,3,4,5$$

$$\sum_i R_{is}^q \leq 1 \qquad \forall s, q$$

$$R_{is}^{q_1} = R_{is}^{q_2} \qquad \forall (q_1, q_2) \text{ through node } k, \forall k \in \delta_i, \forall i$$

$$R_{ij}^q \in \{0,1\} \qquad \forall q, i, j$$

This case example together with the case example of a river basin development problem (details in the next chapter) show the framework developed in the book can tackle the design of flexibility in a broad range of engineering systems.

Chapter 6 Case example: River Basin Development

A hypothetical river basin illustrates the central ideas in this book about real options "in" projects. It involves a set of possible hydropower station sites, reservoir capacities, and installed capacity alternatives. The first phase of options identification uses screening and simulation models to identify real options "in" projects, a subset of projects (with specified locations, reservoir capacities, and installed capacities), for consideration in the real options analysis. The second phase of options analysis addresses the options for timing and choice of projects over 30 years given the uncertain development of energy prices, and of course subject to budget constraints and costs. The issue of whether to build any particular project in a certain period is considered an option – it is a right but not an obligation to build the project in the period. The model for the analysis of these real options readily examines the set of compound options "in" projects. The final products of the analysis include a contingent developing strategy for the river basin development that provides significant improvement in performance (thanks to the use of flexible design and an implementation process that responds to actual situations) and a much-improved valuation of the projects important for investors interested in the projects.

6.1. Case introduction

The case example concerns the development of a hypothetical river basin involving decisions to build dams and hydropower stations. The developmental objective is mainly for hydro-electricity production. Irrigation and other considerations are secondary because the river basin is in a remote and barren place.

The total length of the river is 1570 km. A notable characteristic of the river basin is that it turns 180 degrees around a mountain, forming a 150 km bend-over (See a schematic of the river basin as Figure 6-1). Another important characteristic of the river is that it drops quickly, from the origin at an elevation of 5400 m to the end of 980 m. Total drop is 4420m. The gross theoretical power generation capacity of the river is around 22,000 MW. This hypothetical river is in China, and we use Chinese currency (RMB) to value the projects.

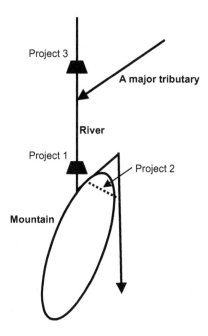

Figure 6-1 Schematic of the hypothetical river basin

There are 3 major projects under consideration for further development. Each project consists of mainly a dam and a hydropower plant. See the schematic of the river as Figure 6-1. The three projects are Project 1, Project 2, and Project 3 as described below.

Project 1

The considered design alternatives for Project 1 are described in Table 6-1:

Table 6-1: Project 1 Design Alternatives

		Alternative 1	Alternative 2	Alternative 3	Alternative 4
Normal Water Storage Level	m	1870	1880	1890	1900
Dead Water Level	m	1790	1800	1810	1820
Reservoir Capacity	Billion m^3	6.97	7.76	8.62	9.64
Adjustable Capacity	Billion m^3	4.52	4.91	5.32	5.75
Installed Power Generation Capacity	MW	3060	3240	3420	3600

The stream flow data for Project 1 is provided in Table 6-2.

Table 6-2: Stream flow for Project 1

Unit: m^3/s	Average	Stream Flow		
		p = 10%	p = 50%	p = 90%
Yearly	1200	1510	1180	904
Dry season (Dec - May)	437	505	435	371

Project 2

As the schematic of the river shows in Figure 6-1, the River has a very big turn-around. Project 1 is built at the beginning of the turn-around. Project 2 will dig an 18 km long tunnel that bypasses 123 km of the river. It also has a much smaller dam compared to

Project 1 or 3. A shallow gradient river section is bypassed and approximately 300m head[1] can be captured.

The choices for the installed capacity are 1,500 MW, 1,600 MW, and 1,700 MW. The choices for tunnel water speed are 4.0m/s, 4.5m/s, and 5.0m/s. The tunnel diameter can be calculated from the tunnel water speed. The choices for tunnel/turbine combination are 3 tunnels 3 turbines, 2 tunnels 4 turbines, and 2 tunnels 6 turbines. The important engineering parameter choices can be shown in Figure 6-2:

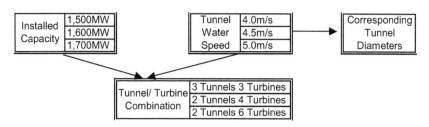

Figure 6-2: Project 2 Tunnel Parameters

Project 3

The design alternatives for Project 3 are shown in Table 6-3. Project 3 hydropower station has adjustable capability over year. Project 3 will add considerable power generating capacity to downstream stations, because of its excellent capability to store water during the wet season and release water during the dry season. See Table 6-4: Project 3 Adding Capacity to Downstream Stations. The stream flow data for Project 3 is provided in Table 6-5.

Since the major purpose of the development of the river basin is for power generation, the key uncertain economic parameter here is the price of electricity, which may vary

---

[1] In hydropower generation, head measures the difference of elevation of water the generator can exploit.

dramatically as China develops economically and moves toward market determination of prices. We should account carefully for this critical uncertain element in planning. If we optimize using an expected electricity price alone, the optimization usually leads to economies of scale arguments indicating that bigger is better. Unfortunately, given the uncertain economic elements, in many cases it does not pay to build as big as possible by exploiting economies of scale, since the demand is often insufficient to justify the big capacity. If the demand is insufficient for a while, it may be difficult for the project to repay loans and it will incur extra loans and extra interest burden. This may be a significant financial loss considering the huge investment nature of water projects. In some cases, it will affect the financial feasibility of a project (see Wang, 2003). It may well be more attractive to build smaller projects with options thinking (Mittal, 2004). This reality is a prime motive for studying real options in large-scale engineering systems.

Table 6-3: Project 3 Design Alternatives

		Alternative 1	Alternative 2
Normal Water Storage Level	m	2840	2880
Dead Water Level	m	2760	2800
Reservoir Capacity	Billion m^3	7.68	12.03
Adjustable Capacity	Billion m^3	5.28	7.49
Installed Capacity	MW	2500	3000

Table 6-4: Project 3 Adding Capacity to Downstream Stations

		Alternative 1	Alternative 2
Increased Dry Season Power Generating Capacity	MW	3109	2841
Increased Yearly Power Generation	TkWh	8.906	10.642

Table 6-5: Stream flow for Project 3

Unit: m³/s	Average	Qp		
		p = 10%	p = 50%	p = 90%
Yearly	657	836	657	502
Dry season (Dec - May)	283	344	283	229

6.2. Options identification

We analyze the case example using the analysis framework developed in Chapter 5. The first step is to identify options started with a screening model.

6.2.1. Screening model

The screening model identifies initial configurations of design parameters for the river basin development, which are sites, reservoir storage capacity, and installed electricity generation capacity. The objective function is to maximize the net present value (NPV) of the river basin development. The constraints include water continuity, reservoir storage capacity, hydropower production, and budget constraints.

The model

Before developing the objective function and constraints for the screening model, first let us define site indices and season indices as in Table 6-6 and Table 6-7.

Table 6-6: Site index "s"

1	2	3
Project 1	Project 2	Project 3

Table 6-7: Season index "t"

1	2
Wet season (June to November)	Dry season (December to May)

Objective Function

The objective function is to maximize net benefit, or

$$\text{Max Benefit} - \text{Cost}$$

$$\text{Benefit} = \sum_s \sum_t \beta^P P_{st}$$

$$\text{Cost} = crf(\sum_s \alpha_s(V_s) + \delta_s(H_s))$$

Where P_{st} is hydroelectric power produced at site s for season t, β^P is the hydropower benefit coefficient, $\alpha_s(V_s)$ is the reservoir cost at site s, $\delta_s(H_s)$ is the power plant cost at site s, crf is the capital recovery factor:

$$crf = \frac{r(1+r)^N}{(1+r)^N - 1}$$

where N is the number of years of life of the project and r is the discount rate.

The reservoir cost coefficient are as follow:

$$\alpha_s(V_s) = FC_s + VC_s \cdot V_s$$

where V_s is the reservoir storage capacity for site s, FC_s is the fixed cost for the reservoir at site s, and VC_s is the variable cost for the reservoir at site s. The cost curves for each dam are estimated on the basis of the studies of Sichuan Hydrology and Hydropower Institute (2002) as described in the section of "estimation of parameters".

The hydropower cost coefficients is as follow

$$\delta(H_s) = \delta_s \cdot H_s$$

where H_s is the installed electricity generation capacity for site s, δs is the variable cost for a power plant at site s. Here a linear relationship between the installed capacity and the cost is assumed. Though it is crude, this linear cost relationship is a good approximation for the purpose of this book.

Constraints

There are four groups of constraints: continuity constraints, reservoir storage and capacity constraints, hydropower constraints, and interdependence constraint.

[Continuity constraints]

Continuity constraints are those constraints ensuring conservation of mass. The water that enters a point on the river stream must leave that point if it has not been stored. Because the evaporation loss is only a very small fraction of the water, evaporation is not considered here.

$$S_{32} - S_{31} = (Q_{in,1} - X_{31})k_t$$

$$S_{31} - S_{32} = (Q_{in,2} - X_{32})k_t$$

$$E_{11} = X_{31} + \Delta F_{31}$$

$$E_{12} = X_{32} + \Delta F_{32}$$

$$S_{12} - S_{11} = (E_{11} - X_{11})k_t$$

$$S_{11} - S_{12} = (E_{12} - X_{12})k_t$$

$$E_{21} \leq X_{11}$$

$$E_{22} \leq X_{12}$$

$$S_{22} - S_{21} = (E_{21} - X_{21})k_t$$

$$S_{21} - S_{22} = (E_{22} - X_{22})k_t$$

where S_{st} is the storage at site s during season t, X_{st} is the average flow from site s for season t, E_{st} is the average flow entering site s for season t, $Q_{in,t}$ is inflow from upstream for season t, k_t is the number of seconds in season t (unit is million seconds), ΔF_{st} is the increment to flow between site s and the next site. Please refer to the schematic of the river basin as Figure 6-3.

The model is run for one typical year with mean water flow for a season, implying all years are the same, the same for each dry season, the same for each wet season. The model sets the initial storage of season 1 equal to the final storage of season 2. The model does not allow variance over years, so it is a deterministic model. The model neither allows us to consider the uncertainty in the water flow nor to study the overyear storage. Once the screening model has determined candidate designs of the system, and reduced the number of variables, the subsequent simulation and timing models introduce the stochastic elements. This strategy allocates computational effort to where it is most productive.

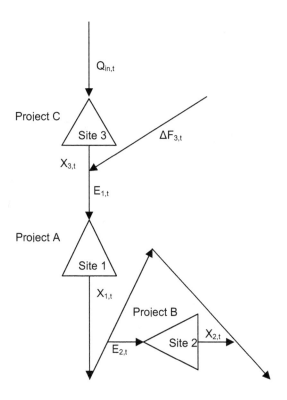

Figure 6-3 Potential projects on the hypothetical river

[Reservoir storage and capacity constraints]

We require that the storage in a reservoir cannot exceed the storage capacity V_s during any season t or at any site s,

$$S_{st} - V_s \leq 0$$

Here a 0-1 integer variable yr_s is introduced. If the integer variable is 0, the dam at site s is not built; if the integer variable is 1, the dam at site s is built. If the dam at site s is not built, the storage capacity at any time at s cannot be bigger than 0; if the dam at site s is built, the storage can be bigger than 0, but

$$V_s - yr_s \cdot CAPD_s \leq 0$$

where CAPDs is the storage capacity of the largest physically feasible dam that can be built. $CAPD_1$ = 9600 (10^6 m^3), $CAPD_2$ = 25 (10^6 m^3), $CAPD_3$ = 12500 (10^6 m^3).

Another constraint is the storage-head relationship needed for hydroelectric production, because the energy produced is proportional to the head. The constraint relates the head, A_{st}, with the storage:

$$S_{st} - \sigma_s (A_{st}) = 0$$

realistically, σ_s should be nonlinear. But such representation also results in non-convex feasible region and local optima, a difficulty. When solving the problem, we should check if we have enough confidence that the result is a global optimum.

[Hydropower constraints]

The production of hydropower obeys relatively straightforward defined physics law and relatively well-defined technical process. Three decision variables are considered in the formulation: the flow, the head, and the installed capacity.

The first hydropower constraint is the physical law for energy conversion:

$$P_{st} - 2.73 \cdot e_s \cdot k_t \cdot X_{st} \cdot A_{st} \leq 0 \qquad \text{Equation 6-1}$$

where P_{st} is the hydroelectric energy produced at site s during season t in MWH, k_t is the number of seconds in season t, e_s is the power plant efficiency factor.

$$k_t = (60Sec / Min) \times (60Min / Hour) \times (24Hour / Day) \times (30Day / Month) \times (6Month / Season)$$
$$= 15.552 \times 10^6 \, Sec / Season$$

e_s is approximately taken to be 0.7.

The unit conversion factor 2.73 is calculated as follows: since 1 Joule = 1 N·m (or $m^2 \cdot kg/s^2$), 1 W = 1 Joule/s, per m^3 of water weighs 10^3 Kg, and per m^3 of water can generate $10^3 \cdot g$ power (where g is the acceleration of gravity, equal to 9.81 m/s^2). One more issue to think about is that in Equation 6-1, the time unit considered in million seconds. So we need a conversion factor:

$$\frac{10^3 \, kg / m^3 \cdot 9.81 m / s^2}{60 \min / hour \cdot 60s / \min} = 2.73 \frac{kg \cdot hour}{m^2 s^3}$$

Let us check the units of this conversion factor. Since e_s has no unit, X_{st} has a unit of m^3/s, A_{st} has a unit of m, k_t has a unit of million s, so the term $2.73 \cdot e_s \cdot k_t \cdot X_{st} \cdot A_{st}$ in Equation 6-1 has a unit of million kg·m^2·hour/s^3, or MWH.

The other variable to be accounted for in the process of energy production is the plant capacity, an upper bound on energy production. h_t is the number of hours in season t, and H_s is the capacity of the power plant in MW. Since the power plant is not going to produce at full capacity all of the time, a factor e_p is defined as the ratio of the average daily production to the daily peak production. It is taken as 0.35 in this study. An implicit assumption in the screening model is that the production pattern does not vary day to day in a season. For a detailed study, we need to include the daily variation in production pattern in the simulation model.

$$P_{st} - e_p h_t H_s \leq 0$$

Solutions of the model frequently showed wide seasonal variation of storage and heads. But marked head variation results in inefficiency. The following three constraints are added to limit the magnitude of head variation (Major and Lenton, 1979)

$$AMIN_s - A_{st} \leq 0$$

$$A_{st} - AMAX_s \leq 0$$

$$AMAX_s - 2AMIN_s \leq 0$$

Another constraint is that the capacity of power plant at site s has an upper bound.

$$H_s - CAPP_s \leq 0$$

According to the data from Sichuan Hydrology and Hydropower Institute, $CAPP_1$ is 3600 MW, $CAPP_2$ is 1700 MW, $CAPP_3$ is 3200 MW.

Estimation of Parameters

This section provides explanations for the methods used to derive the principle economic parameters for the screening model.

The price of electricity is taken to be 0.25 RMB /KWH as of 2001. All the benefit and cost estimates in this study are indexed to 2001 as a base year.

For each of the projects we need to get the following three functions and all their parameters:

Reservoir cost: $\alpha_s(V_s) = FC_s + VC_s \cdot V_s$

Power plant cost: $\delta(H_s) = \delta_s \cdot H_s$

Volume-head curve: $S_{st} = \sigma_s(A_{st})$

All the data and curves are derived or estimated from a report of Sichuan Hydrology and Hydropower Institute (2002).

Project 1

[Reservoir cost]

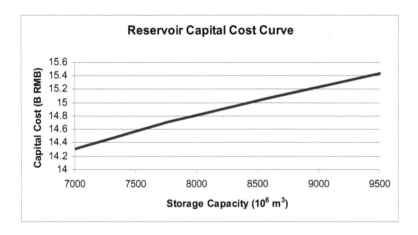

Figure 6-4 Reservoir cost curve for Project 1

Reservoir cost curve for Project 1 is as Figure 6-4, where FC_1 = 11.19 B RMB, VC_1 = 4.49×10⁻⁴ B RMB/10⁶ m³. The parameters are calculated using the least square criteria.

[Power plant cost)
δ_1 is approximately 7.65 ×10⁻⁴ B RMB/MW

[Volume-head curve]

The volume-head curve is as Figure 6-5. The relation between head and volume is approximately expressed by

$$\sigma_1(A_{1t}) = 0.14 A_{1t}^2$$

Note that reservoir volume and head cannot be chosen independently. The relationship between them is decided by the shape of the reservoir.

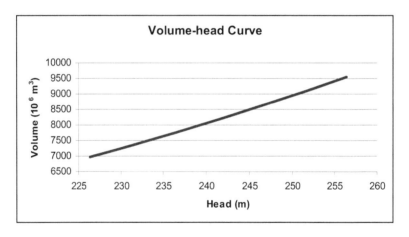

Figure 6-5 Volume-head curve for Project 1

Project 2

The parameters for Project 2 are special, because Project 2 constructs several tunnels to bypass a section of shallow-gradient river.

[Reservoir cost]

The capacity of the reservoir for Project 2 is only around 10 to 20 million m³, the cost of the reservoir is negligible compared to the scale of the tunnel and power plant cost.

[Power plant cost curve]

δ_1 is approximately 1.85 $\times 10^{-3}$ B RMB/MW

[Volume-head curve]

Head is fixed at 280m. Though the head is changing with respect to the capacity of the reservoir, the reservoir is too small and the change is within 10m, the head is assumed to be fixed.

The constraint $S_{2t} - \sigma_s(A_{2t}) = 0$ should be replaced by $A_{2t} = 280$

Project 3

[Reservoir cost]

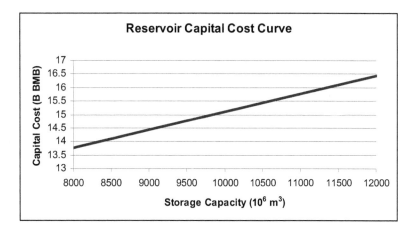

Figure 6-6 Reservoir cost curve for Project 3

The reservoir cost curve for Project 3 is as Figure 6-6, where FC_3 = 8.41 B RMB, VC_3 = 6.68×10^{-4} B RMB/10^6 m^3. The parameters are calculated using the least square criteria.

[Power plant cost]

δ_3 is approximately 8.80 ×10^{-4} B RMB/MW

[Volume-head curve]

The volume-head curve for Project 3 is as Figure 6-7. The relation between head and volume is approximately expressed by

$$\sigma_3(A_{3t}) = 0.15A_{3t}^2$$

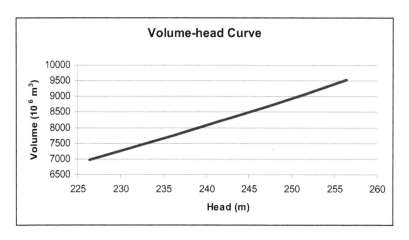

Figure 6-7 Volume-head curve for Project 3

Complete formulation of the screening model

Objective function:

$$\text{Max Benefit} - \text{Cost}$$

$$\text{Benefit} = \sum_s \sum_t \beta^P P_{st}$$

$$\text{Cost} = crf(\sum_s \alpha_s(V_s) + \delta_s(H_s))$$

Continuity constraints:

$$S_{32} - S_{31} = (Q_{in,1} - X_{31})k_t$$

$$S_{31} - S_{32} = (Q_{in,2} - X_{32})k_t$$

$$E_{11} = X_{31} + \Delta F_{31}$$

$$E_{12} = X_{32} + \Delta F_{32}$$

$$S_{12} - S_{11} = (E_{11} - X_{11})k_t$$

$$S_{11} - S_{12} = (E_{12} - X_{12})k_t$$

$$E_{21} \le X_{11}$$

$$E_{22} \le X_{12}$$

$$S_{22} - S_{21} = (E_{21} - X_{21})k_t$$

$$S_{21} - S_{22} = (E_{22} - X_{22})k_t$$

Reservoir storage and capacity constraints:

$$S_{st} - V_s \le 0$$

$$V_s - yr_s \cdot CAPD_s \le 0$$

$$S_{st} - \sigma_s(A_{st}) = 0 \,^1$$

Hydropower constraints:

$$P_{st} - 2.73 \cdot e_s \cdot k_t \cdot X_{st} \cdot A_{st} \leq 0$$

$$P_{st} - e_p h_t H_s \leq 0$$

$$AMIN_s - A_{st} \leq 0$$

$$A_{st} - AMAX_s \leq 0$$

$$AMAX_s - 2AMIN_s \leq 0$$

$$H_s - CAPP_s \leq 0$$

Table 6-8 List of Variables for the screening model

Variable	Definition	Units
yr_s	Integer variable indicating whether or not the reservoir is constructed at site s	
Sst	Storage at site s for season t	$10^6 m^3$
Xst	Average flow from site s for season t	m^3/s
Est	Average flow entering site s for season t	m^3/s
Pst	Hydroelectric power produced at site s for season t	MwH
Ast	Head at site s for season t	m
Hs	Capacity of power plant at site s	MW
Vs	Capacity of reservoir at site s	$10^6 m^3$
$AMAXs$	Maximum head at site s	m
$AMINs$	Minimum head at site s	m

---

[1] Except for Project B, the constraint is replaced by $A_{2t} = 280$.

Table 6-9 List of Parameters for the screening model

Parameter	Definition	Units	Value
$Q_{in,t}$	Upstream inflow for season t	m³/s	(374,283)
$CAPDs$	Maximum feasible storage capacity at site s	10⁶m³	(9600, 25, 12500)
$CAPPs$	Upper bound for power plant capacity at site s	MW	(3600, 1700, 3200)
ΔF_{st}	Increment to flow between sites s and the next site for season t	m³/s	(389, 154) for site 3, others are 0
e_s	Power plant efficiency at site s		0.7
k_t	Number of seconds in a season	Million Seconds	15.552
h_t	Number of hours in a season	Hours	4320
Yst	Power factor at site s for season t		0.35
β^P	Hydropower benefit coefficient	10³ RMB /MWH *	0.25
FCs	Fixed cost for reservoir at site s	B RMB	(11.19,0,8.41)
VCs	Variable cost for reservoir at site s	B RMB/10⁶m³	(4.49×10⁻⁴, 0, 6.68 ×10⁻⁴)
δs	Variable cost for power plant at site s	B RMB/MW	(7.65×10⁻⁴, 1.85×10⁻³, 8.80 ×10⁻⁴)
r	Discount rate		0.086
crf	Capital recovery factor for 60 years		0.087

* All RMB in the study is indexed to 2001 as a base year.

Applying screening model to sort out most interesting flexibility

This example screening model involves 46 variables and 58 constraints. It contains only the most important considerations, yet has a fair amount of technical detail.

After the screening model is established, some detailed consideration and care should be taken when applying it. What uncertain parameters should be examined? What levels of the uncertain variables should be placed into the screening model? After establishing the screening model, we suggest the following steps to use the screening model systematically to search for the interesting real options (flexibility):

Step 1. list key uncertain variables. The uncertain variables could be exogenous or endogenuous. They could be market uncertainty, cost uncertainty, productivity uncertainty, technological uncertainty, etc.

Step 2. find out the standard deviations or volatilities for the uncertain variables. This can be computed by historical data, implied by experts' estimation[1], or using the data from comparable projects.

Step 3. perform sensitivity analysis on the uncertain variables to pick out the several most important uncertain variables for further analysis. Tornado diagram is a useful tool for such sensitivity analysis.

Step 4. list different levels of the important uncertain variables as inputs for the established screening model to identify where the most interesting real options (flexibility) are.

To demonstrate the procedure, we use the case example of the river basin development.

---

[1] The experts give their most likely estimation, pessimistic estimation (better than 5% of cases), and optimistic estimation (better than 95% of cases). Since 5% and 95% represent the range of $\pm 1.65\sigma$, we can estimate σ from the experts' opinion.

Step 1. The important uncertain variables for the river basin development are identified to be the price of electricity, fixed cost of reservoir, uncertain variable cost of reservoir, uncertain variable cost of power plant, and the variable water flow.

Step 2. For the purpose of illustration, we pick out three important uncertain variables for further scrutiny: electricity price, fixed cost of reservoir, and variable water flow. This is only for illustration purpose, and real studies should examine more uncertain variables - the most important uncertain variables may be outside people's expectation or intuition.

The volatility of the electricity price is derived by experts' estimation. The author interviewed two Chinese experts on China energy market in the China Development Bank to get their pessimistic, most likely, and optimistic estimate of the electricity price in three years after 2002. The experts reached the optimistic price estimate of 0.315RMB/kWh and pessimistic estimation of 0.18RMB/kWh both with 95% confidence, which corresponds to a volatility of 6.96% per year (for details see Wang, 2003, pp.101).

The standard deviation of the construction cost is estimated from the standard deviation for the cost of megaprojects (Flyvbjerg, et al., 2003, pp. 16). The standard deviation is 39%.

Table 6-10 Parameters for distribution of water flows (Adjusted for 60-years span)

	Mean	Annual s.d.	60-year s.d
$Q_{in,1,y}$	374	87.7	11.3
$Q_{in,2,y}$	283	45.4	5.9
ΔF_{31y}	389	89.8	11.6
ΔF_{32y}	154	9.7	1.3

The standard deviations of water flow are calculated from historical data. It is listed in Table 6-16. Note, since we consider the project for a life span of 60 years, the annual standard deviation should be translated to a 60-year average standard deviation by dividing the annual standard deviation by $\sqrt{60}$ (see Table 6-10)

Step 3. With the understanding of volatility or standard deviation of the three uncertain parameters. We can draw a tornado diagram regarding the change of net benefit due to one standard deviation/volatility change of one of the important uncertain variables, with other uncertain variables keep at the expected value. To do so, we just change one variable at a time (with 1 standard deviation or volatility) in the screening model, run the optimization and get the corresponding optimal net benefit[1] to be drawn in the tornado diagram. The resulted tornado diagram is as Figure 6-8.

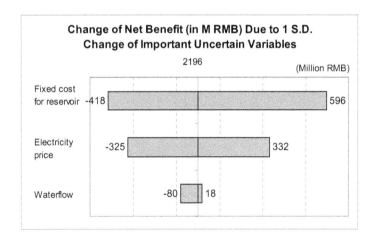

Figure 6-8 Tornado chart for screening model

---

[1] The corresponding optimal design changes given different level of uncertain variables.

From the tornado diagram, we understand the uncertainties on fixed cost of reservoir and electricity price are the most important uncertainties. This book studies in-depth the uncertainty of electricity, however, for the following considerations:

- this study expands the application of options theory to engineering design, so the initial interest leans to that related to financial market. Electricity is market traded and its market risk are relatively well understood;
- costs are more complicated than electricity price. Electricity price is readily market observable, while the fixed cost has a lot of components and hard to know exactly when the project is being constructed. Moreover, the dynamics of uncertainty of reservoir fixed cost is less understood. Despite the importance of study on uncertainty of fixed cost, the purpose of this book is to lay out a general framework for designing flexibility into engineering systems to deal with uncertainties. So we treat the easier electricity uncertainty as a first step to demonstrate the general framework, and call for further work to study the peculiarity of uncertainty of fixed cost in depth.

Step 4. The current electricity price is 0.25 RMB/KWH, but we study the conditions when the electricity price is 0.10, 0.13, 0.16, 0.19, 022, 0.28, 0.31 RMB/KWH. The levels cover most of the range of the experts' pessimistic (0.18 RMB/KWH) to optimistic estimate (0.31 RMB/KMB). To be conservative, we also screen at very low electricity prices as a stress test to see what could happen in the worse case.

Results

Given the 8 levels of electricity price, we get 8 preliminary configurations of the projects. Each configuration has been optimized for a particular price level. The optimization model was written in GAMS© (the GAMS code can be found in Appendix 6A), and the results are shown in Table 6-11. The "Optimal value" represents the maximum net benefit calculated by the objective function. We find that the designs of (H_1 = 3600 MW,

V_1 = 9.6 X 10^9 m^3) and (H_2 = 1700 MW, V_2 = 2.5 X 10^7 m^3) are robust with regard to the uncertainty of the electricity price[1]. But for the project at site 3, the optimal design changes when the price of electricity changes, and this is the place in which we should study further to design flexibility.

Note that for case 1, no projects are built; for cases 2 and 3, site 3 is not built. In a real application considering many more sites, there may be a great number of sites entered the screening models and most of them are screened out. Although the current electricity price is 0.25 RMB/KWH, we cannot assume that the design corresponding to this price is optimal, because the screening model does not consider the uncertainty of electricity price. Follow-on analysis is needed.

Table 6-11: Results from the screening model

Case	Electricity Price (RMB/KWH)	H_1 (MW)	V_1 ($10^6 m^3$)	H_2 (MW)	V_2 ($10^6 m^3$)	H_3 (MW)	V_3 ($10^6 m^3$)	Optimal Value (10^6RMB)
1	0.10	0	0	0	0	0	0	0
2	0.13	3600	9600	1700	25	0	0	367
3	0.16	3600	9600	1700	25	0	0	796
4	0.19	3600	9600	1700	25	1564	6593	853
5	0.22	3600	9600	1700	25	1723	9593	1607
6	0.25	3600	9600	1700	25	1946	12242	2196
7	0.28	3600	9600	1700	25	1966	12500	2796
8	0.31	3600	9600	1700	25	1966	12500	3396

Each design for site 1, 2 and 3 represents an option, though the sources of them are subtly different: all options present timing feature, but only option at site 3 has flexible design feature where we consider different reservoir capacity and power plant capacity

---

[1] The only case that designs of (H_1 = 3600 MW, V_1 = 9.6 X 10^9 m^3) and (H_2 = 1700 MW, V_2 = 2.5 X 10^7 m^3) are not optimal is when the price of electricity is extremely low - where we do the stress test and if is out of the range of experts' estimation.

design. See Table 6-12. Although the features of options have been understood, the final options specification will not be reached until the test of simulation model.

Table 6-12 Sources of options value for designs

	Sources of option value	
	Value of timing	**Value of flexible design**
Design at site 1	Yes	No
Design at site 2	Yes	No
Design at site 3	Yes	Yes

6.2.2. Simulation model

The example simulation model introduces stochastic considerations, both in electricity price and in seasonal flows. The simulation model is operated with 60 years of 6-month seasonal flows. The simulation model takes into account the aspects of over year storage that is hard to consider in the screening model. By using the simulation model, more aspects of the configurations chosen by the screening model are revealed, especially the hydrologic reliability. The simulation model in this book does not look into the hydrologic reliability issue deeply, because the main purpose of the book is about real options "in" projects and the model established here is for illustration of that purpose. A detailed real study, however, should carefully test the plan's hydrologic reliability.

If the time series of the water flow is set to consist of the seasonal means repeating themselves year after year (no shortages with regard to the plan obtained by the screening model) and the price of electricity is not changing, the simulation model should get the same result as of the optimal value of the screening model. But the natural variability of water flow and electricity price will make the result (net benefit) of each run of the simulation different, and the average of the net benefit is not going to be the same as the result from the screening model. The simulated results should be lower because the configurations are not going to benefit from excess water when water is more than the reservoir can store, and losses during droughts will be completely suffered.

Using the simulation model, the several configurations from the screening model are tested. The optimal configuration by the screening model given the electricity price as of the current price of 0.25 RMB/KWH is not necessarily the best configuration after stochasticity of the uncertain parameters is taken into account.

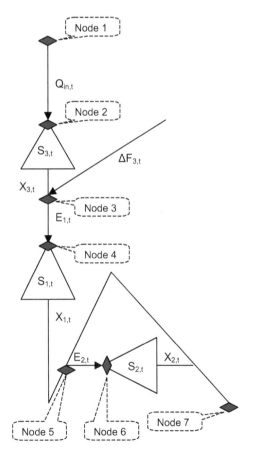

Figure 6-9 River Basin as network of nodes

Table 6-13 List of design variables

Variable	Definition	Unit
H_1	Capacity of power plant at site 1	MW
H_2	Capacity of power plant at site 2	MW
H_3	Capacity of power plant at site 3	MW
V_1	Capacity of reservoir at site 1	$10^6 m^3$
V_2	Capacity of reservoir at site s	$10^6 m^3$
V_3	Capacity of reservoir at site s	$10^6 m^3$

The river basin as a Network of Nodes

In the simulation model, the river basin is described as a network of nodes. The nodes can be categorized as "start node", "reservoir node", "confluence node", and "terminal node" in this case example. Nodes could also include "irrigation node", "import/export node", and "continuation node". Figure 6-9 shows such a network, where node 1 is a start node, node 2 is a reservoir node, node 3 is a confluence node, node 4 is a reservoir node, node 5 is a confluence node even though the water stream is split in two at this point, node 6 is a reservoir node, and node 7 is a terminal node.

Associated with those nodes that represent structural components are design variables. Design variables define system configurations and are chosen as the input to the simulation model by the analyst. The design variables in the simulation model are decided by the results of the screening model, specifically the values of certain decision variables of the screening model. The list of design variables can be found in Table 6-13.

Simulation calculations

Besides the design variables defined in Table 6-13, the other variables and parameters are as the Table 6-14 and Table 6-15.

Basically almost all the relationships in the simulation model are the same as those in the screening model. Only several differences:

- The subscript y in the simulation model is not used in the screening model that optimize over an average year without considering the variation of water and storage over a number of years.
- Several equations are added or modified to take into account the operating rules and average heads. Details follow.

Dynamics of the uncertain inputs

The water inflows are uncertain, including the flow $Q_{in,t,y}$ (Upstream inflow into site 1 for season t for year y) entering the river basin and the increment ΔF_{3ty} (Increment to flow between sites 3 and 1 for season t year y) to flow between site 3 and site 1, as well as the changing electricity power price.

Table 6-14 List of Variables for the simulation model

Variable	Definition	Unit
Ssty	Storage at site s for season t year y	$10^6 m^3$
Xsty	Average flow from site s for season t year y	m^3/s
Esty	Average flow entering site s for season t year y	m^3/s
Psty	Hydroelectric power produced at site s for season t year y	MwH
Asty	Head at site s for season t year y	m

The best fit to the water flow data is a lognormal distribution if the water flow does not dry up during the year. We used river flow data from the National Weather Service (http://waterdata.usgs.gov/nwis/sw). The best fit is the lognormal distribution.

Table 6-15 List of Parameters for the simulation model

Parameter	Definition	Unit	Value
e_s	Power plant efficiency at site s		0.7
T_{st}	Target release at site s for season t	$10^6 m^3$	(4490, 0, 3920)
k_t	Number of seconds in a season	Million Seconds	15.552
h_t	Number of hours in a season	Hours	4320
e_p	Power factor		0.35
β^P	Hydropower benefit coefficient	10^3 RMB /MwH	0.25
FCs	Fixed cost for reservoir at site s	B RMB	(11.19,0,8.41)
VCs	Variable cost for reservoir at site s	B RMB/$10^6 m^3$	(4.49×10^{-4}, 0, 6.68×10^{-4})
δs	Variable cost for power plant at site s	B RMB/MW	(7.65×10^{-4}, 1.85×10^{-3}, 8.80×10^{-4})
r	Discount rate		0.086
crf	Capital recovery factor for 60 years		0.087

The parameters for the lognormal distribution (refer to equation Equation 3-4) of water flows are as Table 6-16:

The price of electricity power price is assumed to follow a Geometric Brownian motion

$$d\beta^P = \mu_e \beta^P dt + \sigma_e \beta^P dz$$

where μ_e is the drift rate for the electricity price, σ_e is the volatility for the electricity price, and dz is a basic Wiener process (See Section 3.2.2.) In this analysis, consistent with the parameters used in the author's master's thesis (Wang, 2003), μ_e is taken to be − 0.33% per year, σ_e is taken to be 6.96%.

Table 6-16 Parameters for the distribution of water flows

	Mean	Standard Deviation
$Q_{in,1,y}$	374	87.7
$Q_{in,2,y}$	283	45.4
ΔF_{31y}	389	89.8
ΔF_{32y}	154	9.7

To set up the simulation, the process is discretized yearly:

$$\frac{\beta_{n+1}^{P} - \beta_{n}^{P}}{\beta_{n}^{P}} = \mu_{e} \cdot \Delta t + \sigma_{e}\varepsilon_{n}$$

where β_{n}^{P} is the price of electricity for year n, Δt here is 1 (year), ε_{n} is a random draw from a standardized normal distribution, $\phi(0,1)$

A sample realization path of the electricity price for 60 years is shown in Figure 6-10. The simulation is built using Crystal Ball©.

For the example analysis, the movement of the electricity price is assumed to follow a geometric Brownian motion (GBM). This is not necessarily the best model for electricity price: a mean-reverting proportional volatility model might improve the quality of analysis (Bodily and Del Buono, 2002). However, GBM is sufficient to illustrate the analysis framework and stochastic mixed-integer programming methodology. To use a different stochastic process, we need to generate an appropriate scenario tree, and the analysis framework remains valid.

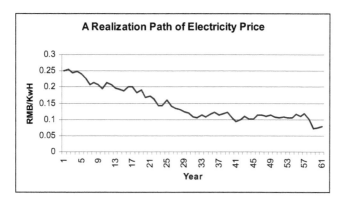

Figure 6-10 A realization path of electricity price

Continuity equations

The flow continuity relationships in the simulation model are the same as those of the screening model. This is true for many of the other relationships as well, such as the hydropower relations. The simulation model evaluates the effect of water flow stochasiticity and electricity uncertainty on the system's net benefits. The simulation model enables the study of the effect of the over-year storage on the system performance, which cannot be evaluated by the screening model. In the screening model, the concept of shortages does not exist - a project is feasible if all the constraints are satisfied, and infeasible if they are not. The concept of shortage, however, is a key part of the simulation model.

The continuity equations are as follows:

For season 1

At node 2: $\quad S_{31,y+1} = (Q_{in,2,y} - X_{32y})k_t + S_{32y}$

At node 3: $\quad E_{11y} = X_{31y} + \Delta F_{31y}$

At node 4: $S_{1,1,y+1} = (E_{12y} - X_{12y})k_t + S_{12y}$

At node 5: $E_{21y} \leq X_{11y}$

At node 6: $S_{2,1,y+1} = (E_{2,2,y+!} - X_{22y})k_t + S_{22y}$

For season 2

At node 2: $S_{32y} = (Q_{in,1,y} - X_{31y})k_t + S_{31y}$

At node 3: $E_{12y} = X_{32y} + \Delta F_{32y}$

At node 4: $S_{12y} = (E_{11y} - X_{11y})k_t + S_{11y}$

At node 5: $E_{22y} \leq X_{12y}$

At node 6: $S_{22y} = (E_{21y} - X_{21y})k_t + S_{21y}$

Operating rules

One of the most important characteristics of the simulation model is the consequence of the consideration of shortages: a distinction must be made between actual releases and target releases. In the simulation model, the target releases must be prespecified and the actual releases are determined during a simulation run by means of an operating rule.

Reservoir releases are determined by operating rules for each reservoir. The basic tradeoff of an operating rule is whether to release water in times of shortage to meet the current downstream demands, or keep some of the water in order to reduce future potential shortages.

This book uses a set of simple operating rules as proposed by Fiering (1967). Although more complicated and better rules can be studied, the rule by Fiering is good enough for the purpose of this research. The actual release X_{st} at site s for season t is a function both of target release T_{st} and of water availability. Three cases are considered (see Figure 6-11):

> Case I: there is not enough water to meet the target release. The reservoir will be emptied in order to meet the downstream demands.

Case II: there is enough water to meet downstream demands. Water exceeding downstream demands is kept in storage.

Case III: the available water minus the downstream demands exceeds maximum storage. All water that cannot be kept must be spilled.

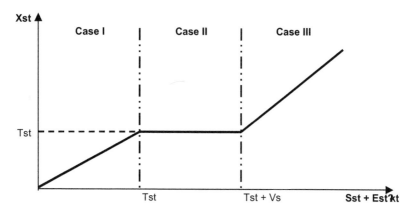

Figure 6-11 Operating rule

The three cases lead to the following equations for X_{st}:

$$
X_{sty} = \begin{cases}
S_{sty}/k_t + E_{sty}; & S_{sty} + E_{sty} \cdot k_t \leq T_{st} \\
T_{st}/k_t; & T_{st} < S_{sty} + E_{sty} \cdot k_t \leq T_{st} + V_s \\
(S_{sty} - V_s)/k_t + E_{sty}; & S_{sty} + E_{sty} \cdot k_t > T_{st} + V_s
\end{cases}
$$

Thus, the storage $S_{s,t+1}$ at the beginning of the following season can be represented by:

$$
S_{s1y} = \begin{cases}
0; & S_{s,2,y-1} + E_{s,2,y-1} \cdot k_t \leq T_{st} \\
S_{s,2,y-1} + E_{s,2,y-1} \cdot k_t - T_{st}; & T_{st} < S_{s,2,y-1} + E_{s,2,y-1} \cdot k_t \leq T_{st} + V_s \\
V_s; & S_{s,2,y-1} + E_{s,2,y-1} \cdot k_t > T_{st} + V_s
\end{cases}
$$

$$S_{s2y} = \begin{cases} 0; & S_{s1y} + E_{s1y} \cdot k_t \leq T_{st} \\ S_{s1y} + E_{s1y} \cdot k_t - T_{st}; & T_{st} < S_{s1y} + E_{s1y} \cdot k_t \leq T_{st} + V_s \\ V_s; & S_{s1y} + E_{s1y} \cdot k_t > T_{st} + V_s \end{cases}$$

The target release in this research is calculated on the basis of a seasonal energy target that is meant to ensure the supply for anticipated energy demand for different seasons. Without losing insights into the real options topic in this thesis, the seasonal energy target is simplified to be a constant for all seasons. The seasonal energy target is set to be 1800 MW, 0 MW, and 1000MW for Project 1, Project 2, and Project 3, respectively. Since the storage capacity for Project 2, V_2, is virtually zero compared to the storage capacities for Project 1 and 3, all the water will be consumed to produce energy without storage at site 2 (Project 2) regardless of the target release or energy target. Given the average heads for Project 1 and Project 3 are taken to be 210m and 240m respectively and equation (A_{st}^* is the average head defined in Equation 6-2 and Equation 6-3)

$$T_{st} = \frac{P_{sty}}{2.73 \cdot e_s \cdot A_{st}^*}$$

the corresponding target releases are 4.49×10^9 m^3 and 3.92×10^9 m^3 for Project 1 and 3, respectively.

Hydropower equations

The total energy produced in season t at site s is calculated by

$$P_{sty} = 2.73 \cdot e_s \cdot k_t \cdot X_{sty} \cdot A_{sty}^*$$

where the average head A_{st}^* is calculated by

$$A_{s1y}^* = \frac{A_{s1y} + A_{s2y}}{2}$$ Equation 6-2

$$A_{s2y}^* = \frac{A_{s2y} + A_{s,1,y+1}}{2}$$ Equation 6-3

Another constraints we need to think about is the plant capacity, an upper bound of the power production:

$$P_{sty} \le e_p h_t H_s$$

where h_t is the number of hours in season t, and Hs is the capacity of power plant in MW, e_p is the ratio of the average daily production to the daily peak production.

Volume-head curves

In order to get the head, we need the information of volume-head curves, they are:

$$S_{1ty} = 0.14 A_{1ty}^2$$

$$A_{2ty} = 280$$

$$S_{3ty} = 0.15 A_{3ty}^2$$

Objective function

The simulation model contains an objective function for analysts to pick the best configurations under test. The form of the objective function is

Net Benefit = Benefit – Cost

where benefit is the discounted cash flow of revenue from the simulated hydropower energy produced (it is assumed that all power produced can be sold), and the cost of building the reservoir and the power plants are given by:

$$\text{Cost} = crf \left(\sum_{s} \alpha_s (V_s) + \delta_s (H_s) \right)$$

where *crf* is the capital recovery factor,

$$crf = \frac{r(1+r)^N}{(1+r)^N - 1}$$

N is the number of years of life of the project. $\alpha_s (V_s)$ and $\delta_s (H_s)$ are the same as defined in the screening model.

Initial conditions of the simulation

Since the screening model assumes an average yearly situation, which implies that the circumstances are under steady state. It does not consider the transition state when a dam has just been built and is storing water. Similarly, the simulation model does not relax the assumption that all projects in a configuration are already built all together (the traditional sequencing model or real options timing model will relax this assumption to study the transient state when projects are built one by one). Also, the initial state of the storage of reservoir is taken to be fully stored.

Simulation model run results

The simulation model tested the 8 configurations chosen by the screening model as shown in Table 6-11. The simulation model was constructed using Excel© and Crystal

Ball©. All designs from the screening model in Table 6-11 were tested. The optimal design predicted by the screening model that corresponds to the current electricity price of 0.25 RMB/KWH is not necessarily the best design after various uncertainties, shortage (droughts) and overyear water storage are taken into account.

For cases 5 and 6 (electricity price = 0.22 RMB/KWH and 0.25 RMB/KWH), the simulation results are as Figure 6-12. The runs for the other cases have lower values than either case 5 or 6. Note that the expected NPV in both cases are substantially below those indicated in Table 6-11 (1138 vs. 1607 for case 5; 1098 vs. 2196 for case 6). As indicated before, this result is not unexpected since the capacity often cannot be fully used due to lower flows. Note, also that the lower capacity design (case 5) provides higher expected NPV than the higher capacity design (case 6) that appeared better in the deterministic design. After considering the price uncertainty and water uncertainty, the seemingly optimal configuration with current electricity price as input performs worse than the configurations with electricity prices lower than actual current price. This is a common, though not necessary, result. This result shows how strong the impact of uncertainty is.

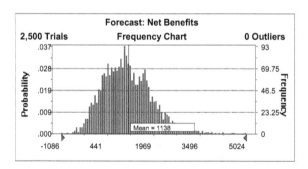

Figure 6-12a: Simulation result for electricity price = 0.22 RMB/KWH (case 5)

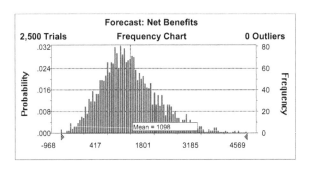

Figure 6-12b: Simulation result for electricity price = 0.25 RMB/KWH (case 6)

Table 6-17 Portfolio of options for water resources case

	Design specifications	Exercise time
Option at Site 1	H_1 = 3600 MW, V_1 = 9.6 X 10^9 m^3	Any time
Option at Site 2	H_2 = 1700 MW, V_2 = 2.5 X 10^7 m^3	Any time
Option at Site 3	One of $\begin{cases} H_3 = 1564 \text{ MW}, V_3 = 6.93 \text{ X } 10^9 \text{ m}^3 \\ H_3 = 1723 \text{ MW}, V_3 = 9.593 \text{ X } 10^9 \text{ m}^3 \\ H_3 = 1946 \text{ MW}, V_3 = 12.242 \text{ X } 10^9 \text{ m}^3 \end{cases}$	Any time

Satisfying the requirements of various technical considerations such as robustness and reliability, the design with the highest expected benefit from the simulation is those corresponding to the electricity of 0.22 RMB/KWH (Case 5). So the timing options for site 1 and 2 are (H_1 = 3600 MW, V_1 = 9.6 X 10^9 m^3) and (H_2 = 1700 MW, V_2 = 2.5 X 10^7 m^3), we have the right to build a project as the specifications, but we do not have the obligation to build them and have the room to observe what happens and decide where to build a project. The option for site 3 contains both timing option and variable design option. We choose 3 design centered Case 5, or (H_3 = 1564 MW, V_3 = 6.93 X 10^9 m^3), (H_3 = 1723 MW, V_3 = 9.593 X 10^9 m^3), or (H_3 = 1946 MW, V_3 = 12.242 X 10^9 m^3). Each design is an option, in that we have the right but not obligation to exercise the option, and the three options are mutually exclusive - only one can be built. A summary for the options sees Table 6-17. This is a portfolio of options. Each option stands for a basic element of flexibility in the project. The next stage will analyze this portfolio of options.

6.3. Options analysis

After successfully identifying the real options "in" projects, we now will analyze the options using the stochastic mixed-integer programming model developed in Chapter 5. The key issues in the real options timing model is the order and timing of the construction of the projects, given various constraints.

6.3.1. Traditional sequencing model

Before establishing the real options timing model, we need to first establish a traditional sequencing model. The real options timing model builds on top of a traditional sequencing model.

From the screening and simulation models, we have obtained a plan of satisfactory design of projects, sizes and locations, under steady state. The next step is to relax the assumption of the previous analysis that all the projects are built at once, and study the transitory construction process from no dams to all planned projects built. The key issue in the question is the order of projects to be built given various constraints in order to maximize the net benefit when taking into account the building process and the water storage from zero to full, as well as the variability of electricity price. The question for the traditional sequencing model is during which time period which project should be built.

The traditional sequencing model assumes that the projects are constructed during 3 time periods, each of equal length of 10 years. The calculation considers 60 years of life for each project (construction time included in the 60 years of life). Different time span can be used, but the difference on results would be small[1].

The traditional sequencing model has almost all 0-1 integer variables except for the flow variables representing stream flow at different points of the river basin and the energy variables representing the energy production, while the screening model has most continuous variables to decide reservoir capacity and power plant sizes that can take any real value within the constraints. In the sequencing model, the sizes of the projects have been decided. The remaining decisions are whether to construct a particular project within a specific period of time. Such decisions are appropriately represented by integer variables.

Constraints of the model

Continuity constraints

The continuity constraints in the sequencing model act as an accounting system for the water use in the river basin. They maintain the links of time and space for water used by different projects, among the most important interdependency between hydropower projects on the river. The continuity constraints in the sequencing model are a little different from what would have been formulated in the screening model: the inclusion of reservoir yield ($\overline{Y_{st}}$, which is obtained from the results of the screening model run) replaces explicit consideration of storage. The continuity constraints are:

---

[1] If we extend the number of years considered, the extra value is very small. For example, if we extend the number of years considered to 70, and assuming every year the benefit is CF and a discount rate of 8.6%. The present value of benefits from year 1 to 60 is $CF \cdot \sum_{j=1}^{60} \frac{1}{(1+0.086)^j} = 11.55CF$,

while the present value of benefits from you 61 to 70 is $CF \cdot \sum_{j=61}^{70} \frac{1}{(1+0.086)^j} = 0.046CF$, or 0.4% of the

value of the first 60 years.

$$X_{i31} = Q_{in,1} + \overline{Y_{31}} \sum_{j=1}^{i} R_{j3} - \overline{c_{31}} R_{i3}$$

$$X_{i32} = Q_{in,2} + \overline{Y_{32}} \sum_{j=1}^{i} R_{j3} - \overline{c_{32}} R_{i3}$$

$$X_{i12} = X_{i32} + \Delta F_{32} + \overline{Y_{12}} \sum_{j=1}^{i} R_{j1} - \overline{c_{12}} R_{i1}$$

$$X_{i21} \leq X_{i11}$$

$$X_{i22} \leq X_{i12}$$

where subscript i denotes the number of the time period, R_{is} denotes if the project at site s is built in period i (1 means built, 0 not). An additional factor that must be taken into account is the filling of a dam when it is just built and flow is required to accumulate storage. The parameter $\overline{c_{st}}$ represents the part of flow to be used during the construction period to ensure a full reservoir of the next period.

Construction constraints

The construction constraints specify the important interdependency between projects besides the continuity constraints. The following construction constraint is introduced to extend the continuity constraint to take into account all periods all together. The constraint guarantees that a project will be constructed at most once (only one term at most can be 1), or not constructed at all (all terms are 0's).

$$\sum_{i} R_{is} \leq 1 \qquad R_{is} \in \{0,1\}$$

Hydropower constraints

The screening model has decided the capacity of the power plant. Although the power generated each year was also an output from the screening model, the result of the screening model cannot be used as a parameter for the sequencing model because, during the construction period, the flow and head are different from the steady state in the screening model. Therefore a hydropower constraint is needed:

$$P_{ist} = 2.73 \cdot e_s \cdot k_t \cdot X_{ist} \cdot \overline{A_{st}} \sum_{j=1}^{i} R_{js}$$

where P_{ist} is the hydroelectric energy produced at site s during season t for time period i in MWH, $\overline{A_{st}}$ is the average head during season t at site s for time period i, k_t is the number of seconds in season t, e_s is the power plant efficiency factor.

Meanwhile, the production of power is capped by the capacity of the plant:

$$P_{ist} - e_p h_t \overline{H_s} \le 0$$

where e_p is defined as the average daily production to the daily peak production, h_t is the number of hours in a season.

Budget constraints

The budget available is the most important scare resource that determines an upper bound on potential projects that can be built in a period. In this case study, it is assumed the budget available will only allow one project built during any period:

$$\sum_{s} R_{is} \le 1$$

Objective function

The objective function is to maximize the net benefit over the planning time horizon. The benefit and cost in a future time period need to be discounted back to present value dependent on how far it is from today.

$$
\text{Max} \quad \sum_i \sum_s \sum_t \beta_P P_{ist} [\sum_{j=1}^{i} R_{js} - (1-f)R_{is}]PV_i + \sum_s \sum_t \sum_i [\beta^P P_{ist} PVO_i R_{is}]
$$
$$
- \sum_i \sum_s \{[\alpha_s(\overline{V_s}) + \delta_s(\overline{H_s})]R_{is} \cdot PVC_i\}
$$

Equation 6-4

When the project is built, water needs to be stored to the specific volume and the water released from the reservoir is different. The term $[\sum_{j=1}^{i} R_{sj} - (1-f)R_{si}]$ will make sure the power production is a fraction of f for the period that the dam is built and needs to be filled up. $\alpha_s(\ )$ and $\delta_s()$ are the same as defined in the screening model.

PV_i is the factor to bring the ten-year annuity of benefit back to the present value as of today.

$$
PV_i = \sum_{j=10(i-1)+1}^{10i} \frac{1}{(1+r)^j}
$$

PVO_i is the factor to bring the annuity after the 3 10-year periods till the end of 60 year life span of a project. If the project is built in the first 10-year period, the value of the project needs to be considered from year 31 to 60; if the project is built in period 2, the value of the project needs to be considered from year 31 to year 70; if the project is built in period 3, the value of the project needs to be considered from year 31 to year 80.

$$
PVO_i = \sum_{j=31}^{60+10(1-1)} \frac{1}{(1+r)^j}
$$

PVC_i is the factor to bring the cost, which is presented as the value at the beginning of the period when it is built, back to the present value as of today. The costs obtained form $\alpha_s(V_s)$ and $\delta_s(H_s)$ are the total present value of costs for the project as of the beginning of the period when the project is constructed. Since the length of the each time period is 10 years, the cost needs to be discounted by a factor of PVC_i because period 1 starts now, period 2 begins at year 11, and period 3 begins at year 21.

$$PVC_i = \frac{1}{(1+r)^{10(i-1)}}$$

Variables and Parameters

Some of the parameters are written with bars over letters. Most of them were decision variables in the screening model (except reservoir yields $\overline{Y_{st}}$) and take the values resulting from the screening model run. As the result of the study of the screening model, the best design configurations are those results optimized for the electricity price of 0.22 RMB/KWh. The value of parameters $\overline{H_s}$, $\overline{V_s}$, $\overline{A_{sti}}$, $\overline{yr_s}$, and $\overline{Y_{st}}$ are obtained from that run. The value can be found in Table 6-19.

How to get $\overline{Y_{st}}$

According to the screening model run, if there is no reservoir at site 3, X'_{31} and X'_{32} are just $Q_{in,1}$ and $Q_{in,2}$, or 374 and 283, respectively; if there is a reservoir at site 3, X_{31} and X_{32} are 310.4 and 346.6, respectively. Therefore, $\overline{Y_{31}}$ and $\overline{Y_{32}}$ are −63.6 and 63.6, respectively.

According to the screening model run, if there is no reservoir at site 1, X'_{11} and X'_{12} are just $Q_{in,1} + \Delta F_{31}$ and $Q_{in,2} + \Delta F_{32}$, or 763 and 437, respectively; if there is a reservoir at site 1, X_{11} and X_{12} are 699.4 and 500.6, respectively. The change of flow after the construction of reservoirs is exactly as $\overline{Y_{31}}$ and $\overline{Y_{32}}$, or -63.6 and 63.6, which is the change caused by the upstream Project 3 at site 3. So $\overline{Y_{11}}$ and $\overline{Y_{12}}$ should be 0.

Since the storage capacity at site 2 is very small compared to those at site 1 and site 3, the storage capacity at site 2 is virtually taken as 0 in the formulation. So $\overline{Y_{21}}$ and $\overline{Y_{22}}$ should be 0 as well.

How to get f

Following is the calculation of parameter f in Equation 6-4. Assuming, during the period when a project is built, the benefit each year follows Figure 6-13. For the first 5 years, there is no power produced, between the 5th and 10th year, the production grows linearly until reaching the full capacity in the 10th year.

So, at the beginning of the period, the value of the benefit flow during the period when the project is constructed, V_t, can be written as:

$$V_t = P_{sti} \sum_{j=5}^{10} \frac{(j-5)}{5(1+r)^j}$$

Equation 6-5

V_t can also be expressed as an annuity fP_{sti}:

$$V_t = \sum_{j=1}^{10} fP_{sti} \cdot \frac{1}{(1+r)^j}$$

Equation 6-6

According to Equation 6-5 and Equation 6-6, f can be expressed as

$$f = \sum_{j=5}^{10} \frac{(j-5)}{5(1+r)^j} / \sum_{j=1}^{10} \frac{1}{(1+r)^j}$$

For r = 8.6%, f = 0.226.

How to get $\overline{c_s}$

The parameter $\overline{c_{st}}$ represents the part of flow to be used during the construction period to ensure a full reservoir of the next period. According to the simulation model, a dam could be filled up within 5 years, because we take that the power benefit beginning after 5 years from the onset of the construction, the water fill-up for a dam has little effect on the energy benefits that we are interested in. We can safely take $\overline{c_s}$ to be 0's and have small impact on the final results.

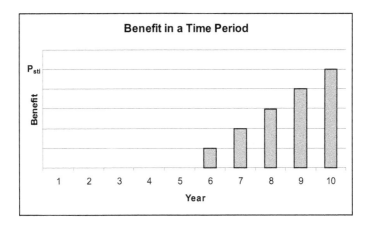

Figure 6-13 Benefit in time periods

The hydropower benefit coefficient – the price of electricity

In the screening model, the electricity price we used is the current price of 0.25 RMB/KWH. However, with the refinement of the simulation, we have understood the uncertainty in the electricity – it follows a Brownian motion with a drift rate μ of -0.33% per year and a volatility σ of 6.96%. In the sequencing model, we have to adjust the hydropower benefit coefficient to reflect this understanding of electricity price dynamics. The expected value of the electricity price after considering its dynamics will be changed to:

$$E(\beta^P) = 0.25 \cdot \frac{\sum_{i=1}^{60}(1-\mu-\sigma^2/2)^i}{60} = 0.21$$

We should use 0.21 RMB/KWH as the hydropower benefit coefficient for the calculation of the sequencing model.

Complete formulation of the traditional sequencing model

Objective function:

$$\text{Max} \quad \sum_i \sum_s \sum_t \beta_P P_{ist} [\sum_{j=1}^{i} R_{js} - (1-f)R_{is}]PV_i + \sum_s \sum_t \sum_i [\beta^P P_{ist} PVO_i R_{is}]$$
$$- \sum_i \sum_s \{[\alpha_s(\overline{V_s}) + \delta_s(\overline{H_s})]R_{is} \cdot PVC_i\}$$

Where

$$PV_i = \sum_{j=10(i-1)+1}^{10i} \frac{1}{(1+r)^j}$$

$$PVO_i = \sum_{j=31}^{60+10(1-1)} \frac{1}{(1+r)^j}$$

$$PVC_i = \frac{1}{(1+r)^{10(i-1)}}$$

Continuity constraints:

$$X_{i31} = Q_{in,1} + \overline{Y_{31}} \sum_{j=1}^{i} R_{j3} - \overline{c_{31}} R_{i3}$$

$$X_{i32} = Q_{in,2} + \overline{Y_{32}} \sum_{j=1}^{i} R_{j3} - \overline{c_{32}} R_{i3}$$

$$X_{i12} = X_{i32} + \Delta F_{32} + \overline{Y_{12}} \sum_{j=1}^{i} R_{j1} - \overline{c_{12}} R_{i1}$$

$$X_{i21} \leq X_{i11}$$

$$X_{i22} \leq X_{i12}$$

Construction constraint:

$$\sum_{i} R_{is} \leq 1$$

Hydropower constraints:

$$P_{ist} = 2.73 \cdot e_s \cdot k_t \cdot X_{ist} \cdot \overline{A_{st}} \sum_{j=1}^{i} R_{js}$$

$$P_{ist} - e_p h_t \overline{H_s} \leq 0$$

Budget constraint:

$$\sum_{s} R_{is} \leq 1$$

Table 6-18 List of variables for the traditional sequencing model

Variable	Definition	Units
X_{ist}	Average flow from site s for season t for time period i	m^3/s
P_{ist}	Hydroelectric power produced at site s for season t for time period I	MwH
R_{is}	0-1 variable indicating whether or not the project is built at site s for time period i	

Table 6-19 List of parameters for the traditional sequencing model

Parameter	Definition	Units	Value
$Q_{in,t}$	Upstream inflow for season t	m³/s	(374,283)
ΔF_{st}	Increment to flow between sites s and the next site for season t	m³/s	(389, 154) for site 3, others are 0
e_s	Power plant efficiency at site s		0.7
k_t	Number of seconds in a season	Million Seconds	15.552
h_t	Number of hours in a season	Hours	4320
e_p	Power factor		0.35
β^P	Hydropower benefit coefficient	10^3 RMB /MwH *	0.25
FC_s	Fixed cost for reservoir at site s	B RMB	(11.19,0,8.41)
VC_s	Variable cost for reservoir at site s	B RMB/10^6m³	(4.49×10^{-4}, 0, 6.68 ×10^{-4})
$\overline{H}_s$	Capacity of power plant at site s	MW	(3600, 1700, 1723)
$\overline{V}_s$	Capacity of reservoir at site s	10^6m³	(9600, 0, 9593)
$\overline{A}_{st}$	Head at site s for season t	M	(262, 262; 280, 280; 240, 253)
$\overline{Y}_{st}$	Reservoir yield at site s for season t (the change of the flow if a reservoir is built)	m³/s	(0, 0; 0, 0; -63.6, 63.6)
$\overline{c}_{st}$	Part of flow at site s season t to be used in the construction period to ensure a full reservoir of the next period	m³/s	0
f	The ratio of average yearly power production during the construction period over the normal production level		0.226

δs	Variable cost for power plant at site s	B RMB/MW	$(7.65\times10^{-4},$ $1.85\times10^{-3},$ $8.80\times10^{-4})$
PVi	Factor to bring 10-year annuity of benefit back to the present value as of year 0 (now)		$(6.532,\ 2.863,$ $1.254)$
$PVOi$	Factor to bring the annuity from year 31 to year 70 back to year 0		$(0.896,\ 0.943,$ $0.963)$
$PVCi$	Factor to bring cost in the ith period back to year 0		$(1,\qquad 0.438,$ $0.192)$
r	Discount rate		0.086
$\overline{yr_s}$	Indicating whether or not the project is built at site s		$(1, 1, 1)$
crf	Capital recovery factor		0.087

Results

Using GAMS (code see Appendix 6B), we get the result is

$R_{11} = 0$, $R_{21} = 0$, $R_{31} = 0$
$R_{12} = 1$, $R_{22} = 0$, $R_{32} = 0$
$R_{13} = 0$, $R_{23} = 0$, $R_{33} = 0$

The optimal net benefit is 3861 Million RMB.

The result means that we are only going to build Project 2 in the first period and do not build Project 1 and 3. The result is different from those of the screening model or the simulation model. It is understandable because the sequencing model takes into account the transition process: reservoirs take time to build and to be filled up; power plants take time to build and gradually increase the production to full capacity. Unfortunately, this transition process is close to present and has a huge impact on the NPV. For the screening model and simulation model, steady state is considered and the huge impact of the transition process is ignored. Using the traditional sequencing model taking into account the transition process (without thinking of flexiblity in the future), we find only one project is worthwhile to build.

If we force to build all projects by demanding $\sum_i R_{is} = 1$, the result is we should build Project 2 in the first period, Project 1 in the second period, and Project 3 in the third period. But the optimal net benefit is 2008.58 Million RMB, about half of the plan that we only build Project 2 in the first period.

Are we really going to build only Project 2 and discard all the other projects? No! Remember that the power price is uncertain and may rise above a point so that it is worthwhile to build one or more other projects. The next section will study how to build price uncertainty systematically into the traditional sequencing model and solve the real options timing model.

Even though we only get the result to build Project 2 in period 1 with this traditional sequencing model, it is a useful model. It gives us an immediate action plan to build Project 2. We will have time to wait for more information to plan sensibly for other projects. Even if we have a good plan for future projects, we still need to adjust the future projects based on how reality unfolds. Any plan for longer future should be dealt with more caution. If there is time to wait for more information and there is no opportunity cost for waiting, early commitment is not only unnecessary but also unwise.

As a baseline, we consider the traditional sequencing model where the electricity price is deterministic, in other words, we do not consider the real options. The sequencing model takes into account the transition process that projects, once built, gradually increase the production to full capacity. It recognizes that time is needed to build reservoirs and power plants and to fill up reservoirs. This deferral of benefits over a number of years has a huge impact on the NPV for large, capital-intensive projects. Thus, projects that appeared good in the screening or simulation model analyses may turn out to be less attractive when sequencing issues are considered.

In short, the above traditional sequencing model has already given us the important information about which project to build first, and further improvement of the sequencing model will be found in the next chapter.

6.3.2. Real options timing model

The real options timing model is the combination of the traditional sequencing model and real options tree lattice algorithm. It is a stochastic mixed-integer programming formulation. We will study the key interesting aspect in this study compared to the previous study, that is, we are going to consider the hydropower benefit coefficient (energy price) uncertainty in the future. And the contingency plan given the realization of actual energy prices will also result from the model. So, the result from the model is a dynamic plan as well as the optimized objective function value.

Path-dependency and breaking of the recombination structure of the binomial tree

The electricity price movements can be described in the tree as Figure 6-14, where β^P is the electricity price, u is the up factor, and d is the down factor.

But the key difference of the formulation of the real options timing model based on this binomial tree and that of financial option is the path-dependent/path-independent feature. In the binomial tree for financial options, the valuation of an option on each node is path-independent, which means it does not matter how the price moves into that state. For example, at year 20, if the price is $\beta^P ud$, the valuation of the option on that node is the same for any path leading into that state, wether β^P -> $\beta^P u$ -> $\beta^P ud$ or β^P -> $\beta^P d$ -> $\beta^P ud$. In the real options timing model based on the tree, the path-independency feature does not hold any more. It does matter, at node $\beta^P ud$, how the price evolved in the past. If it went through $\beta^P u$, a dam might have been built, and that dam will be producing power and generating electricity during the period; however, if it went through $\beta^P d$, no dam might have been built, the expected revenue in that period being different.

So, in the real options timing model, we do need to consider all the paths that lead into final states as in Table 6-20. Where p is the risk-neutral probability. It is defined by:

$$p = \frac{e^{\mu \cdot \Delta T} - e^{-\sigma \sqrt{\Delta T}}}{e^{\sigma \sqrt{\Delta T}} - e^{-\sigma \sqrt{\Delta T}}}$$

where μ is the drift rate, ΔT is the time interval between periods. In the river basin development case example, μ is –0.33% per year, and ΔT is 10 years.

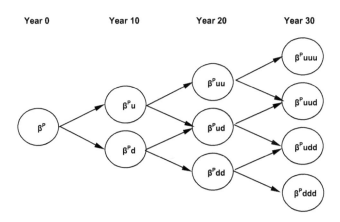

Figure 6-14 Electricity price movement (with recombination)

Table 6-20: Paths of electricity price movement

	Year 0	Year 10	Year 20	Year 30	Probability of path
Path 1	β^P	$\beta^P u$	$\beta^P uu$	$\beta^P uuu$	p^3
Path 2	β^P	$\beta^P u$	$\beta^P uu$	$\beta^P uud$	$p^2(1-p)$
Path 3	β^P	$\beta^P u$	$\beta^P ud$	$\beta^P udu$	$p^2(1-p)$
Path 4	β^P	$\beta^P u$	$\beta^P ud$	$\beta^P udd$	$p(1-p)^2$
Path 5	β^P	$\beta^P d$	$\beta^P du$	$\beta^P duu$	$p^2(1-p)$
Path 6	β^P	$\beta^P d$	$\beta^P du$	$\beta^P dud$	$p(1-p)^2$
Path 7	β^P	$\beta^P d$	$\beta^P dd$	$\beta^P ddu$	$p(1-p)^2$
Path 8	β^P	$\beta^P d$	$\beta^P dd$	$\beta^P ddd$	$(1-p)^3$

Formulation

The real options timing model incorporates uncertainty in the hydropower benefit coefficient (energy price). It gives a contingency plan in reaction to the actual realization of energy price. In this connection, we would again like to point out the path-dependent feature of this problem. Refer to Figure 6-14. For example, in the second stage, if electricity price goes up, a project is built; but if it goes down, no project is built. According to the formula for binomial tree, the middle point of the third stage has a price of $\beta^P ud$ or $\beta^P du$, numerically the same, however, it is different for the following two paths leading into the point because of the hydrological conditions: first path, the price goes up in the second stage, and goes down the third stage with a project changing the water flow; the second path, the price goes down in the second stage, and goes up in the third stage with no project and the natural water flow.

The key differences of the definition of variables/parameters of this sequencing model with the traditional model are as follow: R_{is}, 0-1 variable indicating whether or not the project is built at site s for time period i, is replaced by R_{is}^q, a new 0-1 integer variable for which the subscript s and i have the same meaning and q represents the scenario index from 1 to 8. The hydropower benefit coefficient, β^P, is not a constant any more. It will fluctuate according to the time period i and scenario index q, or it can be written as β^q_{Pi}.

Objective function

Compared to the traditional sequencing model without real options considerations, the objective of the real options timing model is changed to:

$$
\text{Max} \quad
\begin{aligned}
& \sum_s \sum_q p^q \sum_i \sum_t \beta^q_{Pi} P_{ist} [\sum_{j=1}^{i} R^q_{js} - (1-f)R^q_{is}]PV_i + \sum_s \sum_q \sum_i \sum_t p^q \beta^q_{PT} P_{ist} PVO_i R^q_{is} \\
& - \sum_s \sum_q \sum_i p^q \{[\alpha_s(\overline{V_s}) + \delta_s(\overline{H_s})]R^q_{is} \cdot PVC_i\}
\end{aligned}
$$

where

$$
PV_i = \sum_{j=10(i-1)+1}^{10i} \frac{1}{(1+r)^j}
$$

$$
PVO_i = \sum_{j=31}^{60+10(1-1)} \frac{1}{(1+r)^j}
$$

$$
PVC_i = \frac{1}{(1+r)^{10(i-1)}}
$$

Technical constraints

The technical constraints are the same as the traditional sequencing model, including continuity constraints, construction constraint, hydropower constraints, hydropower constraints, and budget constraint. The only difference is due to the change of specification of R_{is} into R^q_{is}. Actually, we can view each path of the electricity price as a run for the traditional sequencing model.

Real options constraints

Moreover, the constraints differ because this formulation adds the real options constraints:

$$\sum_i R_{is}^q \le 1 \qquad \forall s, q$$

$$R_{is}^{q_1} = R_{is}^{q_2} \quad \forall (q_1, q_2) \text{ through node } k, \forall k \in \delta_i, \forall i = 1,...,n \qquad \text{Equation 6-7}$$

Specifically, in our case example, Equation 6-7 are as follow: For i = 1, there is only one decision (refer to Figure 6-15, at year 0, the decision maker has only the information up to that time, and cannot distiguish which path the electricity price will follow), or

$$R_{1s}^q = R_{1s}^1, \quad \forall q \ne 1$$

For i = 2, there are two different decisions (refer to Figure 6-15), or

$$R_{2s}^1 = R_{2s}^2 = R_{2s}^3 = R_{2s}^4$$
$$R_{2s}^5 = R_{2s}^6 = R_{2s}^7 = R_{2s}^8$$

For i = 3, there are 4 different decisions (refer to Figure 6-15), or

$$R_{3s}^1 = R_{3s}^2$$
$$R_{3s}^3 = R_{3s}^4$$
$$R_{3s}^5 = R_{3s}^6$$
$$R_{3s}^7 = R_{3s}^8$$

Our planning horizon is three stages (30 years) though each project is assumed to have a life time of 60 years. At the last time stage, there is no decision (refer to Figure 6-15) and the realization of electricity price is used to calculate the net benefit of projects.

Year 0	Year 10	Year 20	Year 30
β^P	$\beta^P u$	$\beta^P uu$	$\beta^P uuu$
β^P	$\beta^P u$	$\beta^P uu$	$\beta^P uud$
β^P	$\beta^P u$	$\beta^P ud$	$\beta^P udu$
β^P	$\beta^P u$	$\beta^P ud$	$\beta^P udd$
β^P	$\beta^P d$	$\beta^P du$	$\beta^P duu$
β^P	$\beta^P d$	$\beta^P du$	$\beta^P dud$
β^P	$\beta^P d$	$\beta^P dd$	$\beta^P ddu$
β^P	$\beta^P d$	$\beta^P dd$	$\beta^P ddd$

| ↑ | ↑ | ↑ | ↑ |
| 1 Decision | 2 Different Decisions | 4 Different Decisions | No decision |

Figure 6-15 Illustration of Nonanticipativity Constraints in the River Basin Case

Complete formulation of the problem

Objective function:

$$
\text{Max} \quad \sum_s \sum_q p^q \sum_i \sum_t \beta^q_{Pi} P_{ist} [\sum_{j=1}^{i} R^q_{js} - (1-f) R^q_{is}] PV_i + \sum_s \sum_q \sum_i \sum_t p^q \beta^q_{PT} P_{ist} PVO_i R^q_{is}
$$
$$
- \sum_s \sum_q \sum_i p^q \{ [\alpha_s(\overline{V_s}) + \delta_s(\overline{H_s})] R^q_{is} \cdot PVC_i \}
$$

where

$$
PV_i = \sum_{j=10(i-1)+1}^{10i} \frac{1}{(1+r)^j}
$$

$$PVO_i = \sum_{j=31}^{60+10(1-1)} \frac{1}{(1+r)^j}$$

$$PVC_i = \frac{1}{(1+r)^{10(i-1)}}$$

Constraints:

Technical constraints:

$$X_{i31}^q = Q_{in,1} + \overline{Y_{31}} \sum_{j=1}^{i} R_{j3}^q - \overline{c_{31}} R_{i3}^q$$

$$X_{i32}^q = Q_{in,2} + \overline{Y_{32}} \sum_{j=1}^{i} R_{j3}^q - \overline{c_{32}} R_{i3}^q$$

$$X_{i12}^q = X_{i32}^q + \Delta F_{32} + \overline{Y_{12}} \sum_{j=1}^{i} R_{j1}^q - \overline{c_{12}} R_{i1}^q$$

$$X_{i21}^q \leq X_{i11}^q$$

$$X_{i22}^q \leq X_{i12}^q$$

$$P_{ist}^q = 2.73 \cdot e_s \cdot k_t \cdot X_{ist}^q \cdot \overline{A_{st}} \sum_{j=1}^{i} R_{js}^q$$

$$P_{ist}^q - e_p h_t \overline{H_s} \leq 0$$

$$\sum_s R_{is}^q \leq 1$$

Real options constraints:

$$\sum_i R_{is}^q \leq 1$$

$$R_{1s}^q = R_{1s}^1, \ \forall q \neq 1$$

$$R_{2s}^1 = R_{2s}^2 = R_{2s}^3 = R_{2s}^4$$

$$R_{2s}^5 = R_{2s}^6 = R_{2s}^7 = R_{2s}^8$$

$$R_{3s}^1 = R_{3s}^2$$

$$R_{3s}^3 = R_{3s}^4$$

$$R_{3s}^5 = R_{3s}^6$$

$$R_{3s}^7 = R_{3s}^8$$

Table 6-21 List of variables for the real options timing model

Variable	Definition	Units
X_{ist}^q	Average flow from site s for season t for time period i for scenario q	m^3/s
P_{ist}^q	Hydroelectric power produced at site s for season t for time period i for scenario q	MwH
R_{is}^q	0-1 variable indicating whether or not the project is built at site s for time period i for scenario q	

Table 6-22 List of parameters for the real options timing model

Parameter	Definition	Units	Value
$Q_{in,t}$	Upstream inflow for season t	m^3/s	(374,283)
ΔF_{st}	Increment to flow between sites s and the next site for season t	m^3/s	(389, 154) for site 3, others are 0
e_s	Power plant efficiency at site s		0.7
k_t	Number of seconds in a season	Million Seconds	15.552
h_t	Number of hours in a season	Hours	4320
e_p	Power factor		0.35
β_{Pi}^q	Hydropower benefit coefficient for period i and scenario q	10^3 RMB /MwH	

FCs	Fixed cost for reservoir at site s	B RMB	(11.19,0,8.41)
VCs	Variable cost for reservoir at site s	B RMB/10^6m^3	(4.49×10^{-4}, 0, 6.68 ×10^{-4})
$\overline{H_s}$	Capacity of power plant at site s	MW	(3600, 1700, 1723)
$\overline{V_s}$	Capacity of reservoir at site s	10^6m^3	(9600, 0, 9593)
$\overline{A_{st}}$	Head at site s for season t	m	(262, 262; 280, 280; 240, 253)
$\overline{Y_{st}}$	Reservoir yield at site s for season t (the change of the flow if a reservoir is built)	m^3/s	(0, 0; 0, 0; -63.6, 63.6)
$\overline{c_{st}}$	Part of flow at site s season t to be used in the construction period to ensure a full reservoir of the next period	m^3/s	0
f	The ratio of average yearly power production during the construction period over the normal production level		0.226
δs	Variable cost for power plant at site s	B RMB/MW	(7.65×10^{-4}, 1.85×10^{-3}, 8.80 ×10^{-4})
PVi	Factor to bring 10-year annuity of benefit back to the present value as of year 0 (now)		(6.532, 2.863, 1.254)
$PVOi$	Factor to bring the annuity after the 3 10-year periods till the end of 60 year life span of a project		(0.896, 0.943, 0.963)
$PVCi$	Factor to bring cost in the ith period back to year 0		(1, 0.438, 0.192)
r	Discount rate		0.086

$\overline{yr_s}$	Indicating whether or not the project is built at site s		(1, 1, 1)
crf	Capital recovery factor		0.087
σ	Volatility for power price		
ΔT	Number of years between two periods		Year

List of variables and parameters are repeated three times for the screening model, the simulation model, and the real options timing model. Note the subtle differences of variables and parameters in the three models. Each model studies the same problem with different focus, and thus the variables and parameters may be defined in a slightly different way. Such differences are critical and reflect the purposes that each model serves.

6.3.3. Options analysis results

Using GAMS©, we obtain the results for the real options timing model (code see Appendix 6C) as Table 6-23. For example, for the first scenario q = 1 that occurs with probability = 0.138: the electricity prices for the first, second, and third 10-year time period (i = 1, 2, and 3) are 0.250, 0.312, and 0.388 RMB/KWH, respectively. The real options decision variables for Project 2 in the first period and Project 1 in the third period are 1's, and the other 7 real options decision variables are 0's (we have 9 real options decision variables for each scenario, 3 projects times 3 periods each). Therefore, for scenario 1, the decision is to build Project 2 in the first period and Project 1 in the third period. The rest of Table 6-23 can be read in the same way. In summary, as in Figure 6-16, the optimal strategy or contingency plan is to build Project 2 in the first time stage whatever the electricity price is. And build nothing in the second stage. In the last stage, we only build Project 1 in the case that price is up for the second stage and up again for the third stage, for other cases, we build nothing.

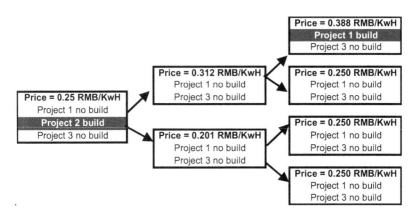

Figure 6-16: Contingency plan

Table 6-23: Results for real options "in" projects timing model (Current electricity price 0.25 RMB/KWH)

Scenario	Electricity Price Realization				Decision			
	$i = 1$	$i = 2$	$i = 3$	Prob		$i = 1$	$i = 2$	$i = 3$
$q = 1$	0.250	0.312	0.388	0.138	Project 1	0	0	1
					Project 2	1	0	0
					Project 3	0	0	0
$q = 2$	0.250	0.312	0.250	0.233	Project 1	0	0	0
					Project 2	1	0	0
					Project 3	0	0	0
$q = 3$	0.250	0.201	0.250	0.233	Project 1	0	0	0
					Project 2	1	0	0
					Project 3	0	0	0
$q = 4$	0.250	0.201	0.161	0.395	Project 1	0	0	0
					Project 2	1	0	0
					Project 3	0	0	0

The overall expected net benefit is 4345 million RMB. As a comparison, the traditional sequening model suggests that only Project 2 should be built in the first time period and that Projects 1 and 3 should never be built, with an overall expected net benefit of 3861 million RMB. Real options add additional value by building Project 1 in favorable situation (electricity price high) and avoiding it in an unfavorable situation (electricity price low). Refer to Figure 6-16. In this example, the added value is not enormous, but the principle is established.

There are three important notes regarding the results:
- This real options timing model provides a contingency plan (as Figure 6-16) depending on how events unfold, as well as the value of projects with real options. Using the real options timing model, we learn to build Project 2 in the first stage, and build Project 1 in the third stage given certain electricity price condition; if using the traditional sequencing model, the decision is to build Project 2 in the first stage, and then build nothing else, surely missing something compared to the real options timing model.
- This contingency plan takes into account the complex interdependencies among projects, in this case, through the water flows (for example, one dam in the upstream will store water and help downstream stations to produce more in dry season). Using conventional options analysis, it is hard to deal with such interdependencies. This example is simpler than real water resources planning; nevertheless, we can use exactly the same methodology, with more computation and other resources, to tackle much more complex real water resources planning problem.
- The value of options is the difference between the optimal benefits from the real options timing model and the traditional sequencing model, 484 million RMB, or 12.5% of the result of traditional sequencing model without real options consideration. Note the valuation of real options "in" projects looks not for an exact numeric result as valuation of financial options, but assesses whether flexible designs are worthwhile. This process about real options valuation is more

about the process of designing flexibility itself than a specific value of optimal benefit.

Finally, a few words about the computational costs: for the example river basin development problem, the number of variables is 187, of which 36 are 0-1 discrete variables, and the number of constraints is 261. It takes a laptop (PIII 650, 192M RAM) less than 2 seconds to use GAMS© to figure out a solution.

6.3.4. Extension of real options timing model to consider multiple designs at a site

In the previous analysis, we have an implicit assumption that all designs are fixed now. Actually we do not do this in reality, and the design will change with regard to how reality unfolds. Considering this aspect of flexibility, we should further add value to the real options timing model developed in the previous section.

Before we start studying multiple designs. We want to see a case that exhibits more optionality. For the current price of 0.25 RMB/KWH, it is too low to have Project 3 entering the picture, while the designs of Project 1 and 2 do not vary a lot with respect to electricity price as shown in Table 6-11.

The case with current electricity price of 0.30 RMB/KWH

The optionality of the plan can be more significant if the current electricity price is higher. If we take the current electricity price as 0.30 RMB, the result is as Table 6-24. The optimal expected net benefit is 6899 million RMB. The corresponding optimal expected net benefit is 5195 million RMB for the traditional sequencing model with a current electricity price of 0.30 RMB/KWH. The options value is 1704 Million RMB, or 32.8% of the result of the traditional sequencing model without real options consideration. The contingency plan is to build Project 2 in the first stage, build Project 1 in the second stage if the price goes up, build Project 3 in the third stage if the price goes up both in the second and third stage. Refer to Figure 6-17. Note the path-dependent feature: in the third stage, for the same electricity price of 0.30 RMB/KWH, Project 1 can have been built or not, depending whether the electricity price in the second stage was high or low.

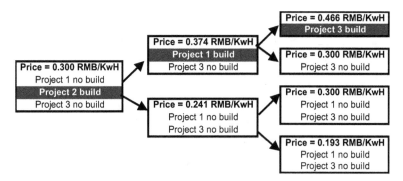

Figure 6-17: Contingency plan (if current electricity price = 0.30 RMB/KWH)

Table 6-24: Results for real options "in" projects timing model (Current electricity price taken to be 0.30 RMB/KWH)

Scenario	Electricity Price Realization				Decision			
	i = 1	i = 2	i = 3	Prob		i = 1	i = 2	i = 3
q = 1	0.300	0.374	0.466	0.138	Project 1	0	1	0
					Project 2	1	0	0
					Project 3	0	0	1
q = 2	0.300	0.374	0.300	0.233	Project 1	0	1	0
					Project 2	1	0	0
					Project 3	0	0	0
q = 3	0.300	0.241	0.300	0.233	Project 1	0	0	0
					Project 2	1	0	0
					Project 3	0	0	0
q = 4	0.300	0.241	0.193	0.395	Project 1	0	0	0
					Project 2	1	0	0
					Project 3	0	0	0

Multiple designs for Project 3 when current electricity price is 0.30 RMB/KWH

To verify this idea, we study multiple designs for Project 3. As Table 6-11 shows, the design of Project 3 varies significantly with regard to the power price, and this is the place of focus point for designing flexibility. We are now considering adjusting the design for Project 3 according to the unfolded electricity price, as well as the decision whether to build Project 3 based on the electricity price.

We designated 3 sets of potential designs from Table 6-11 corresponding to the cases that the electricity prices are 0.22, 0.25, 0.28 RMB/KWH, respectively. And the optimization can choose among the sets to find the optimal one. The 0-1 variable $z(n)$ defines which design is chosen, if set n is chosen then $z(n) = 1$, otherwise $z(n) = 0$. Since only one set can be chosen, so

$$\sum_n z(n) = 1$$

The 3 sets of design for the case are as Table 6-25. Corresponding to each design, the water flow parameters are different as Table 6-26.

Table 6-25 Three sets of design for multiple height choice

	Design 1	Design 2	Design 3
H1	3600	3600	3600
H2	1700	1700	1700
H3	1723	1946	1966
V1	9600	9600	9600
V2	25	25	25
V3	9593	12242	12500

Table 6-26 Different water flow corresponding to different height

	Design 1	Design 2	Design 3
$\overline{Y_{11}}$	262	262	262
$\overline{Y_{12}}$	262	262	262
$\overline{Y_{21}}$	280	280	280
$\overline{Y_{22}}$	280	280	280
$\overline{Y_{31}}$	240	274	277
$\overline{Y_{32}}$	253	286	289

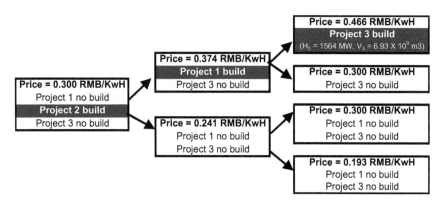

Figure 6-18 Contingency plan for current electricity price of 0.30 KWH/RMB

Using GAMS to solve the formulation (GAMS code is as Appendix 6D). The optimal benefit raised to 7129 million RMB, compared to the result of 6899 million RMB using only the design of $H_3 = 1723$ and $V_3 = 9593$. Compared with the result of the traditional sequencing model without real options consideration, 5195 Million RMB, the options value is 1934 Million RMB, or 37.2% of the result of the traditional sequencing model without real options consideration. The new result is to go with Project 2 in the first period, and go with Project 1 in the second period if price goes up, and go with Project 3 with the design of H_3 = 1966 and V_3 = 12500 if the price goes up again in the third period. In this way, we further expand the power of the method to consider more flexibility. See the contingency plan as Figure 6-18.

Chapter 7 Policy Implications –

Options in Practice

Before studying how to apply options thinking in practice, we have to first make sure that policy makers would use it. This means that options analysis has to make a substantial improvement over previous analysis tools. Two substantial improvements must be presented to and understood by policy makers. The two improvements from options thinking are value of options (better mean) and/or change of distribution (desirable distribution)

Better mean: It is often the case[1], once options are imbedded in the systems planning process, that there is an improvement in the value of the project. We have estimated the options value for case example of river basin is 484 million RMB as described in Section 6.3.3. The value is substantial when uncertainty is higher, which is measured by a higher volatility.

Desirable distribution: With options on hand, the distribution of the value of a project is changed. Some extreme values are avoided for the downside, and/or some more favorable upside potential is obtained. Smaller downside risk is very important for many policy makers. The more risk averse a policy maker, the more important a smaller downside risk is. For some cases, even if the mean of the value of a project is smaller than another project, the project may be still be preferred by policy makers because of a smaller downside risk. An options analysis will provide a plan with a smaller downside and/or bigger upside. The case example of parking garage development (Section 4.3.4.) provides a good illustration of manipulation of distribution by options, especially the VAR curves in Figure 4-5.

---

[1] Under some circumstances, people may use an option that actually decreases the mean, e.g. insurance.

After showing policy makers of the significance of options thinking, we are going to further clarify the relationship between flexibility and economies of scale, and then some application tactics is very important in making real impact on options thinking. Meanwhile, we should be aware that options analysis is not a panacea, sometimes it works, sometime it does not. So it is important to understand the uncertain reality and choice of appropriate methodology.

7.1. Flexibility vs. Economies of Scale

Big engineering projects often exhibit significant economies of scale. Economies of scale are common in industries where cost largely depends on the envelope to the structure (a quantity expressed in terms of the square of the linear measure), and capacity depends on the volume (a quantity expressed in terms of the cube of the linear measure). In such situations, total cost grows approximately to the 2/3 power of capacity.(de Neufville, 1990; Chenery, 1952). The desire to take advantage of economies of scale has been a prime motivation for building large facilities such as ships, aircraft, chemical plants, thermal power stations, and many other kinds of manufacturing. Most pertinently for civil engineering, it characterizes tunnels, pipelines, dams, hydro plants, and water treatment facilities (de Neufville, 1970). A typical deterministic design practice forecasts expected values of uncertain parameters, and uses those expected values as inputs for further analysis and design. Optimization with such design parameters often leads to economies of scale.

However, when plants are only operating at some fraction of full capacity, larger and supposedly more economical facilities may in fact be relatively uneconomical, compared to smaller facilities, when capacity is not fully used. For example, in our case example of river basin development, the actual benefit from the simulation model is lower than what the screening model suggests. This is because the projects are not going to benefit from excess water when water is more than the reservoir can store, but on the downside, they are fully exposed to drought conditions. Thus occasional high levels of water do not compensate for lost revenues due to occasional low levels of water. Due to these uncertainties, the economies of scale seemingly apparent under the deterministic screening model are reduced.

An important trade-off exists between economies of scale and flexibility. The economies of scale are only fully achieved when demand catches up to the size of the facility. Meanwhile, the facility has to be built and paid for in advance, and interest or other opportunity costs are incurred. Larger facilities offer more economies of scale, but also need more time to achieve them. Smaller facilities offer more flexibility to meet the actual realization of demand with less unused capacity, and need less time and lower cost to build. There is an economically optimum size of construction for each phase of expansion. As Manne (1961, 1967) has shown, this sweet spot can be calculated precisely – assuming that one knows the economies of scale factor, the discount rate over the period, and the rate of growth of demand. All his studies were based on a deterministic view. If uncertainties regarding the demand and supply are high, the rules Manne developed may be misleading.

Only exceptionally do long-term forecasts actually hit the mark. As a general rule, we can expect that the actual long-term future demand will be different from what was projected as the most likely scenario. Designers must expect that big engineering systems will have to serve any one of a range of possibilities.

The fact that there is great uncertainty about the future loads has two implications for design, as Mittal (2004) has documented:

- the optimal size of the design should be smaller than that defined by Manne (1961, 1967), that is, planned for a shorter time horizon; yet
- the design should be easy to adjust for the range of possible long-term futures, or in other words, flexible.

Note the two design implications have important implications on Economies of Scale: basically, one of the statements brings forward the idea that optimal size may be one planned for a shorter time horizon, thus we are implicitly saying Economies of Scale may not be that "real".

In short, the reality of great long-term uncertainty means that system designers need to adopt a strategy of development whose evolution depends on the future as it unfolds. Specifically, they should insure their project against downturns and utilize upturns potential in anticipated requirements (by building smaller facilities and thus paying an insurance premium by sacrificing potential economies of scale), while designing in the capability to expand aggressively if future growth demands larger capacity. In other words, system designers need to build in "real options".

7.2. Making Real Impact of the Options Thinking

The author spent 4 months in the China Development Bank in Beijing to help the bank study the real options method and implement real options thinking in their project evaluation process. The China Development Bank is the major investor of China public projects in energy, transportation, and other public areas where other investors are unwilling or unable to invest.

To use options thinking in a developing country like China, there are more difficulties than using the methodology in a developed country like the US. In the US, there are still many problems need to be solved before real options method can be applied maturely and widely. The major difficulties of the real options method in developing countries are how to make decision-makers understand the method and obtain the quality data necessary for options valuation models.

The options methodology develops a proactive and analytical view of uncertainty and flexibility. Despite the difficulties mentioned above, it has a good possibility to make a real impact in a developing country such as China. Besides the methodology itself, the organizational tactics and skills to make people accept new ideas are also among the keys to make real impact of the method.

This section discusses the two aspects of promoting the real options method in China, and more generally, in developing countries. The first aspect is the intrinsic difficulties in promoting real options method in developing countries, and the second is the organizational wisdom to advocate new thinking.

7.2.1. Intrinsic Difficulties in Promoting Real Options methodology

First of all, it should be stressed the real options method has a lot of room to grow in developing countries, despite the difficulties discussed in this section. In other words, these difficulties do not make the method inapplicable in developing countries.

Understanding the method

The options theories and models always seem arcane to people, the partial differential equations... the dynamic processes... people usually think options are the work of rocket scientists.

In China, what makes things even worse is that, since China is afraid that the trading of options is easy to foster fraudulence in its weak legal system, financial options are not traded in local financial markets.. So Chinese people do not know what options are intuitively like people in developed countries.

If a person does not establish an understanding of financial options, he/she will find it very hard to develop the idea of what the real options are and to have confidence in the method. For example, in the China Development Bank where the author worked, a manager heard that the real options method overvalues projects, then he had deep misunderstanding and mistrust of the method because the real options method seems like a black box and hard to comprehend.

After all, a manager is not going use a method if he/she does not understand it.

Data

Options analysis needs a lot of historical data to do objective analysis. In developed countries, the abundant historical data on financial market provides the power of options analysis - there is little subjective element in the analysis, and the magic of market tells all. However, in developing countries, there is no complete financial market, and consequently, the data is incomplete. Even with some data, the financial market is decided by too much government interference so the information is highly distorted. China is such an example. So, even if there is data, people must be very cautious when using it. Also, because China has been undergoing tremendous reform for a long time, there has been several major system changes in the past years. The state changes imply that the available data does not reflect the current situation.

A way to circumvent the data problem is to use simulation to obtain important parameters such as the volatility. In this way, the model risk is huge because the model includes of subjective assumptions of the model-makers.

Although the difficulties in helping managers understand the methodology and the availability of data problems, the real options method will be able to spread fast in developing countries because of its insights into uncertainty and flexibility. The thinking is invaluable, nevertheless.

7.2.2. Organizational Wisdom to Advocate New Thinking

In this section, the author's personal experience in specific settings is introduced. Readers please judge if it works in the specific organization and specific country you are working in.

During the work in the China Development Bank, the author realized the tremendous difficulties in promoting new thinking, especially in some big institutions that have a rigid bureaucracy and highly risk-averse. After working in the bank for months, the author figured out some tactics to quicken the process of adoption of new ideas. Although any good ideas will be used in the end, the process is very slow and painful. For example, it took 20 years for people to use the NPV method widely. However, a smart way will make the process of adoption of real options methodology faster.

Humans do not like changes. They would like to follow the old way that they are familiar with. Because of the efforts and pains to, unless people do not face threats or other pressures, they will not take the efforts to change. Change is difficult and we must be patient and know how to make things happen tactically.

After working in the China Development Bank. The author understands the following 4 elements are the keys for the success of letting people accept the options thinking. They are

- support from the top management;
- identify the needs of the organization;
- from simple to complex;
- perfect communication with grass-root people in the organization.

The support from the top management is key to success. Basically, in most big organizations, one of the most important motivations of people is promotion. If the top management would like to promote something, people will have enough motivation to take the endeavor to study the new ideas. When the author did the real options study and promotion in the China Development Bank, one of the top managers gave me full support on the project.

Another important tactic is to identify the needs of the organization, and firstly show them that the method can be used to deal with the questions that they are most interested in. A bank does not care too much about the value of a project, because what they can get at most is the principal plus interest. If the project earns more, or has more upside, it has nothing to do with the bank. On the other hand, the bank bears the risk of losing the big chunk of the principal (while the interest they can earn is only a small part compared to the principal). So they are more interested in the risk management rather than valuation. So when the author presented the real options to the Bank during the first several weeks, the author did not mention many classical benefits of the options valuation, such as valuation of the flexibility. Instead, the author used simple simulation showing the idea of right not obligation of options can be designed into a contract and reduce the default risk. By focusing on the needs of the bank rather than the traditional stressed benefits of real options, people in the Bank are getting interested in the method and understand the method gradually.

The real options method is not a simple method, though it looks transparent to experts who study the method everyday. It does take a while for those that have never heard about it to understand the method. So it is important to show the thinking and method from simple to complex step by step. In the bank, when the author showed people there how to evaluate the options, firstly the author did not show them the profound options valuation formula or even binomial tree. the author just showed them the idea of cut probability density function to value an option which is much more intuitive with the presentation of a software like Crystal ball.

A perfect communication with the bank people is the key to success, especially the grass-roots people. Their trust in the author is one of the necessary conditions to success. In the bank, the author had to do a successful case to show the method workable. Though the author know the real options method, the author does not know the intricacy of project evaluation. For example, the author did a project valuation on a hydropower project study, but the author is not an expert on the China hydropower industry. So my real options analysis might have made no sense to the experts in the bank who have studied the industry for decades. So the author has to make sure that my reports were collaborated with experts in the bank. In the bank, the author made two very good friends, one is an expert in China energy industry and another is the people who the bank designated to cooperate with me. With friendly and perfect communication with these two grass-roots people, the author was able to understand the bank faster and avoid many stupid mistakes.

7.3. Uncertain Reality and Appropriate Method

Nature is inherently unknown. People can handle the situation with known risks with known distributions, but people just cannot deal with the situations where the uncertain factor is unknown, or the uncertain factor is known but the distribution of the uncertain factor is unknown. Anticipatable risks should be managed, but less can be done about other system risks. When people have to enter into the domains that are ultimately unknown, they need to invest in information to the best capacity possible. If you have done what you can do and still fail in the end, no one is to blame.

Several common quantitative tools for decision making are listed below:

- Discounted cash flow (DCF)
- Decision tree analysis (DTA)
- Real Options

A useful classification of uncertainty helps people to understand the environment better, and identify the best quantitative tools for decision making to deal with the specific uncertainty. And most importantly, people need to understand that the right decision and method do not necessarily lead to a happy outcome.

Suppose the outcome, denoted Π, is a function of N influence parameters:

$$\Pi = \Pi(w_1, w_2, ..., w_N)$$

Usually, the project payoff Π cannot be predicted with certainty.

Complexity: Many of the parameters are not random, and they could be under the influence of the decision-makers – for example, engineering designs. Interdependencies among influence parameters may make the optimization a significant task. Thus, complexity is the centerpiece of much of deterministic analysis tools, such as the NPV method and Linear Programming. Complexity increases with the number of nonseparable variables in Π. There is no uncertainty involved in complexity, actually. Another example of complexity is scheduling classes in a big university.

Risk: Some influence parameters are not deterministic but dynamic. There is payoff variation of Π associated with those parameters. There are two kinds of situation: the distributions of an influence parameter can be known or unknown. If the distribution is known, and the influence parameter is market-traded assets, the standard real options approaches apply, such as the Black-Scholes formula, binomial tree, and simulation methods. If the distribution is known, and the influence parameter is not marketed traded, people may use the decision tree analysis and binomial tree to design contingent plan and value the project. However, if the influence parameter is known while the distribution is unknown, both decision analysis and real options have no ground to work, and the focus should be put on understanding the distribution of the influence parameter. Or a decision tree analysis with the probability as an unknown parameter x is a possible method. Sensitivity analysis with respect the unknown parameter x can be done to get important insights into the problem.

Ambiguity: The decision-makers may not be aware of some the influence parameters at all. The decision-makers explicitly ignore the payoff impact of those influence parameters, but implicitly takes "default" value for those influence parameters. The past is extrapolated subject to changes in known parameters in the category of complexity and risk. For any real decision, it is hard to avoid ambiguity, or in other words, not to miss any influence variable. This is the key reason why forecast is almost certainly to be wrong. For the case in an ambiguous environment, the best method is to invest in information carefully, learning by trial and error. It is like entering a room without light to find a way out. After bumping onto the table, the pain tells you that there is a table... In such situation, people should avoid betting too much and do not put all the eggs in a single basket.

The above analysis of uncertainty types can be summarized in Table 7-1.

Table 7-1 Uncertain Reality and Appropriate Method

	Complexity	Risk			Ambiguity
		Distribution Known		Distribution Unknown	
		underlyingmarket traded	underlying not market traded		
Methods	Deterministic Valuation (NPV, LP, etc)	Financial Options valuation formula	DTA	DTA (with probability as an unknown to do sensitivity analysis)	Trial and error
		Binomial Tree	Simulation		
		Simulation			
Examples	Scheduling classes in a university	Energy Project valuation with underlying of energy price	River basin development with waterflow distribution known	Whether China energy market will be market allocation or government allocation regime	New technology aimed at unkonwn market

With the understanding of uncertainty types, decision-makers can

- Decide when different analysis techniques are most suitable, when to use real options method, when to apply decision tree analysis, and when to employ the mathematical programming method;

- Understand that there are cases where the analysis techniques do not help much if a lot of ambiguity exists, and the best strategy is to carefully learn by trial and error, and do not make life-or-death bet;

Because of the ambiguity is ubiquitous in real life, sometimes a decision has been optimized given the information available, but the results may not be happy. However. In such case, decision-makers should understand the decision *is* correct and not blame the methodology, and gather information for future decisions.

Chapter 8 Summary and Conclusion

In uncertain environments, flexibility has great value for project design, which usually means to design smaller and design spare features initially, wait for further information, and prepare for different kinds of happenings. This book proposes a comprehensive framework to identify and deal with strategic flexibility as an integral part of project design.

8.1. Background

Options concept is a way to define the basic element of flexibility. The concept of options is a right, but not obligation, to do something for a certain cost within or at a specific period of time. This concept models flexibility as an asymmetric right and obligation structure for a cost within a time frame. This is a basic structure of human decision making – taking advantage of upside potential or opportunities and avoiding downside risks. We can construct complex flexibility using the basic unit of options.

In financial terms, a typical example for options is a "call option" that gives the holder the right to buy an underlying asset for a specified exercise price within or at a specified time. The holder of the call option will exercise this right only if the value of the asset rises above this price, but not otherwise. The key property of an option is the asymmetry of the payoff, an option holder can avoid downside risks and limit the loss to the price of getting the option, while she can take advantage of the upside risks and the possible gain is unlimited. Usually, there is a price or cost to obtain an "option". The groundbreaking Black-Scholes formula is the first work to value a financial option and its developers – Black, Scholes, and Merton – won Nobel Prize in 1997 for the achievement.

"Real options" are termed to emphasize those options involve real activities or real commodities, as opposed to purely financial commodities, as in the case, for instance, of stock options. For example, they may be associated with the valuation of an offshore oilfield, the development of a new drug, the timing of the construction of a highway, or the design of a satellite communications system. Since Myers (1984) coined the term of "real options", researchers have been developing a variety of means to evaluate real options (see for example: Dixit and Pindyck, 1994; Trigeorgis, 1996; Luenberger, 1998; Copeland and Antikarov, 2001).

The common techniques to value real options are: the Black-Scholes formula, simulation, and binomial tree. The Black-Scholes formula is the solution to a differential equation, the major assumptions to derive the differential equation are no-arbitrage and geometric Brownian motion of the price of underlying asset. With the value of parameters, we can use the Black-Scholes to calculate the option value. However, if lack of understanding of the underlying assumptions for Black-Scholes formula, it is very easy to apply the formula blindly and obtain a useless and misleadingly precise "value of real options". Monte Carlo simulation does not have as many assumptions as the Black-Scholes formula. If it is possible to specify the stochastic processes for the underlying uncertainties, and to describe the function between the input uncertain variables and the output payoff, computers can do the "brute force" work. Plausibly, simulation can obtain any valuation that Black-Scholes can get at any specified level of accuracy, and it can tackle problems with complex and non-standard payoffs that Black-Scholes cannot deal with. Binomial tree provides the basis for a dynamic programming algorithm. It is based on a simple representation of the evolution of the value of the underlying asset of an option, and calculates by rolling back from the last time stage to the first stage to get the value of the option. The approach allows the recombination of states to decrease the computational burden. It can also represent various stochastic processes of the underlying asset and various options exercise conditions, so it is powerful and flexible.

8.2. Real options "in" projects

Real options can be categorized as those that are either "on" or "in" projects (de Neufville, 2002). Real options "on" projects are financial options taken on technical things, treating technology itself as a 'black box'". Most classical studies on real options are for those "on" projects, for example, a real investment opportunity in a gold mine (Luenberger, 1998) or investment in capacity expansion for a petroleum chemical company (Amram and Kulatilaka, 1999). Real options "in" projects are options created by changing the actual design of the technical system. Following are three examples of real options "in" projects:

Example 1: "Bridge in bridge"

The design of the original bridge over the Tagus River at Lisbon provides a good example of a real option "in" projects. The original designers built the bridge stronger than originally needed, strong enough so that it could carry a second level, in case that was ever desired. The second level of bridge was an option "in" projects, and the Portuguese government exercised the option in the mid 1990s, building on a second deck for a suburban railroad line (Gesner and Jardim, 1998).

Example 2: Satellite systems

In the late 1980s, Motorola and Qualcomm planned the Iridium and Globalstar systems to serve their best estimates of the future demand for space-based telephone services. Their forecasts were wrong by an order of magnitude (in particular because land-based cell phones became the dominant technology). The companies were unable to adjust their systems to the actual situation as it developed and lost almost all their investments - - 5 and 3.5 billion dollars respectively. However, if the companies had designed evolutionary configurations that had the capability to expand capacity, it would have been possible both to increase the expected value of the system by around 25%, as well as to cut the maximum losses by about 60% (de Weck et al, 2004). Such evolutionary configurations can be realized by designing real options for the room of future capacity expansion. For example, a smaller system with smaller capacity can be established first. For a smaller system, there could be fewer satellites with a higher orbit. One possible real option is to carry extra fuel on each satellite. When demand proves big, the satellites can move to lower orbits with the existing orbital maneuvering system (OMS). With additional satellites launched to lower orbits, a bigger system is accomplished to serve the big demand. The extra fuel carried in the satellites are real options. They can be exercised when the circumstances turn favorable. There is cost to acquire such real options – the cost of designing larger tanks and launching extra fuel. Decision makers have the right to exercise the options, but not the obligation – they can leave the extra fuel on board.

Example 3: Parking garage design

This example is extrapolated from the Bluewater development in England of a multi-level parking garage. A car-parking garage for a commercial center is planned in a region that is growing as population expands. Economic analysis recognizes that actual demand is uncertain, given the long time horizon. If the owners design a big parking garage, there is a possibility that the demand will be smaller and the cost of a big garage cannot be recovered; however, if the owners design a small parking garage, they may miss the opportunity if the demand grows rapidly. To deal with this dilemma, the owners can design a real option into the design by strengthening the footings and columns of the original building so that they can add additional levels of parking easily. This premium is the price to get the real option for future expansion, a right but not an obligation to do so. (de Neufville, Scholtes, and Wang, 2005)

Note the difference between real options "in" projects and the engineering concept of "redundancy". Both real options "in" projects and redundancy refer to the idea that some components should not have been designed if the design were optimized given the assumption that things are not going to change. Redundancy refers to more than enough design elements to serve the same function, while real options "in" projects may not serve the same functions as some currently existing components (though such real options do not prove necessary given the current situation).

Comparison of real options "on" and "in" projects

Real options "on" projects are mostly concerned with the valuation of investment opportunities, while real options "in" projects are mostly concerned with design of flexibility. Some classic cases of real options "on" projects are on valuation of oil fields, mines, and pharmaceutical research projects, where the key question is to value such projects and decide if it is worthwhile to invest in them. The examples of real options "in" projects are extra small rockets on satellites, strengthened footings and columns of a multi-level parking garage, or "bridge in bridge".

Real options "on" projects are mostly concerned with an accurate value to assist sound investment decisions, while real options "in" projects are mostly concerned with "go" or "no go" decisions and an accurate value is less important. For real options "on" projects, analysts need to get the value of options, but for real options "in" projects, analysts do not have to provide the exact value of the options but simply provide what real options (flexibility) to design into the physical systems.

Real options "on" projects are relatively easy to define (a categorization of real options can be found in Trigeorgis, 1993), while real options "in" projects are difficult to define in physical systems. For an engineering system, there are a great number of design variables, and each design variable can lead to real options "in" projects. It is hard to find out where the flexibility can be and where is the most worthy place to design real options "in" project. Identification of options is an important issue for real options "in" projects.

Real options "on" projects do not require knowledge of technological issues, and interdependency/path-dependency is not frequently a salient issue. However, real options "in" projects need careful consideration of technological issues. Complex technological constraints often lead to complex interdependency/path-dependency among projects. Table 8-1 summarizes the comparison between real options "on" and "in" projects.

Real options "on" projects	Real options "in" projects
Value opportunities	Design flexibility
Valuation important	Decision important (go or no go)
Relatively easy to define	Difficult to define
Interdependency/Path-dependency less an issue	Interdependency/Path-dependency an important issue

Table 8-1 Comparison between real options "on" and "in" projects

There is much less literature on real options "in" projects than that on real options "on" projects. Zhao and Tseng (2003) discussed the value of flexibility in multistory parking garages. Zhao, Sundararajan, and Tseng (2004) presented a multistage stochastic model for decision making in highway development that incorporating real options in both development and operation phase. Leviakangas and Lahesmaa (2002) discussed the application of real options in evaluation of intelligent transportation system and pointed out the shortcoming of traditional cost-benefit analysis that may discard the value of real options. Kumar (1995) presented the real options approach to value expansion flexibility and illustrated its use through an example on flexible manufacturing systems. Ford, Lander, and Voyer (2002) proposed a real options approach for proactively using strategic flexibility to recognize and capture project values hidden in dynamic uncertainties. de Neufville, Scholtes, and Wang (2005) developed a spreadsheet Monte

Carlo simulation model to value real options in the design of a multistory park garage and gained insights into real options "in" projects, especially the key trait of real options taking advantage of upside potential while cutting downside risk.

Real options "in" projects are those that are most interesting to project designers. In general, real options "in" projects require a deep understanding of technology. Because such knowledge is not readily available among options analysts, there have so far been few analyses of real options "in" projects, despite the important opportunities available in this field.

There are two key difficulties facing the analyses of real options "in" projects. The first is that while financial options are well-defined contracts and real options "on" projects are easy to construct via different financial arrangements, it is much harder to identify real options "in" projects where there are myriads of design variables and parameters. The second difficulty is that real options "in" projects often exhibit complex path-dependency/interdependency that standard options theory does not deal with. Real options "in" projects are different and need an appropriate analysis framework - existing options analysis has to adapt to the special features of real options "in" projects.

8.3. Analysis framework for real options "in" projects

The analysis of real options "in" projects is in part inspired by the standard procedures for water resources planning described by Major and Lenton (1979). These embody a series of models to generate satisfactory solution for the plan. Because of the size of the problem, in terms of number of parameters and uncertain variables, a single model giving the optimal solution is too complex to establish. So people divide the modeling into a series of models and get a satisfactory solution rather than search the best plan among all possible plans. As Herbert Simon (1957) pointed out: because of the astronomical amount of extrinsic information and human's limited intrinsic information process capacity, in real decision making, people do not search for an optimal decision, instead, they stop looking for better decisions once reaching a satisfactory decision. The process - that divides the decision process into several consecutive models and search for a satisfactory solution rather than an optimal solution - is in accordance with the nature of human decision-making in a very complex and uncertain environment. Specifically, the standard water resources planning procedures divide the process into a:

- deterministic screening model that identifies the possible elements of the system that seem most desirable;
- simulation model that explores the performance of candidate designs under stochastic loads; and
- timing model that defines an optimal sequence of projects.

The process of analysis for real options "in" projects modifies these traditional elements. At a higher level, it divides the analysis into 2 phases as indicated in Figure 8-1: options identification and options analysis.

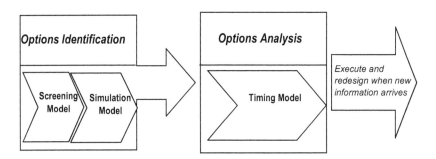

Figure 8-1: Process for Analysis of Real Options "in" Projects

8.3.1. Options identification

For real options "in" projects, the first task is to define the options. This is in contrast with financial options, whose terms (exercise price, expiration day, and type such European, American or Asian) are clearly defined. For real options "in" projects, it is not obvious how to decide their exercise price, expiration day, current price, or even to identify the options themselves. A project design involves a great many decision variables, such as the date to build, capacity, and location, etc. Each of these could place options in. The question is: which options are most important and justify the resources needed for further study? it is only possible, after the options have been identified, to analyze the options to show their value and develop a contingency plan for the management of the projects. This first task for real options "in" projects is not trivial.

Screening model

The screening model is a simplified, conceptual, low-fidelity model for the system. Without losing the most important issues, it can be easily run many times to explore an issue, while the full, complete, high-fidelity model is hard to establish and costly to run many times. The screening model is established to finding out most important variables and interesting real options. From another perspective to look at the screening model, we can think it as the first step of a process to cut small the design space of the system. The design space is extremely big and the possibility for future realization of exogenous uncertain factors is also extremely big. Therefore, we cut design space smaller and smaller in steps, rather than using a holistic model to accomplish all the results in one run.

Specifically, a screening model can be a linear (or nonlinear) programming model:

$$\text{Max: } \sum_j (\beta_j Y_j - c_j Y_j) \qquad \text{Equation 8-1}$$

$$\text{s.t.} \quad \mathbf{TY} \geq \mathbf{t} \qquad \text{Equation 8-2}$$

$$\mathbf{EY} \geq \mathbf{e} \qquad \text{Equation 8-3}$$

Y_j are the design parameters. The objective function (Equation 8-1) calculates the net benefit, or the difference between the benefits and costs, where β_j and c_j are the benefit and cost coefficients. Usually we measure benefits in money terms, though sometimes we do so in other measures, e.g. species saved, people employed, etc. Constraints (Equation 8-2) and (Equation 8-3) represent technical and economic limits on the engineering systems, respectively. Note when feedback exists in the system, the screening model has to carefully take care of the feedback; otherwise, it may produce misleading or erroneous conclusions.

Any parameter in the formulation could be uncertain. There are economic uncertainties in $\mathbf{E}$, $\mathbf{e}$, β_j, or c_j and technical uncertainties in $\mathbf{T}$ or $\mathbf{t}$. After identifying the uncertain variables, we perform sensitivity analysis on those uncertain variables to pick out several most important uncertain variables for further analysis, and then run the screening model with inputs of a range of values for the most important underlying uncertain variables in a systematic way, and then compare the resulting sets of projects that constitute optimal designs for each set of inputs used. If the resulted design for a facility does not change with respect to the uncertain variables, the design for this facility is robust with respect to the uncertain variables (note such design still could represent timing options if we consider the sequence of implementation); however, if the resulted design varies considerably with respect to different levels of the uncertain variables, this design can be the focus point to place design options.

The resulted options have two sources of flexibility value:

- Value of timing. Some part of the project may be deferred. These represent timing options. Its implementation depends on the realized uncertain variables. Since it can catch upside of the uncertainties by implementation, and avoid downside of the uncertainties by holding implementation. Such timing options have significant value by themselves.

- Value of flexible design. Some part of the project may present distinct designs given various realization of uncertainties, compared to the timing options whose design are the same whenever they are built.

An option can contain one or both sources of flexibility value.

The screening model does not consider all the complexities of the system; it considers a large number of possibilities, screens out most of them, and focuses attention on the promising designs. With such simplifications, we are able to focus our attention on identifying the most interesting flexibility, and leave the scrutiny of other aspects to the following models and studies.

Simulation model

The simulation model tests several candidate designs from runs of the screening model. It is a high fidelity model. Its main purpose is to examine, under technical and economic uncertainties, the robustness and reliability of the designs, as well as their expected benefits from the designs. Such extensive testing is hard to do using the screening model. After using the simulation model, we find the most satisfactory candidate design with parameters $(\overline{Y_1}, \overline{Y_2}, ..., \overline{Y_j})$ as well as the real options in preparation for the next stage of options analysis.

8.3.2. Options analysis

After identifying the most promising real options "in" projects, designers need a model that enables them to value the portfolio of options and develop a contingency strategy for their exercise. In contrast to standard financial options analysis, more characteristics are required for the analysis of real options "in" projects, such as technical details and interdependency/path-dependency among options.

The standard binomial options valuation model is based on a simple representation of the evolution of the value of the underlying asset of an option. In each time period, the underlying asset can take one of the two values Su or Sd determined by the volatility of the underlying asset, S is the current asset price, and u and d are the up and down factors respectively. For many periods, the binomial tree is shown in Figure 8-2.

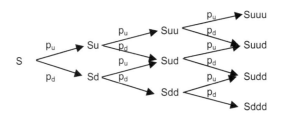

Figure 8-2 Binomial tree

The up factor (u), down factor (d), move-up probability (p_u) and move-down probability (p_d) are decided by:

$$u = e^{\sigma\sqrt{\Delta t}}$$

Equation 8-4

$$d = e^{-\sigma\sqrt{\Delta t}}$$

Equation 8-5

$$p_u = \frac{e^{rT} - d}{u - d}$$

Equation 8-6

$$p_d = 1 - p_u$$

Equation 8-7

where r is the drift rate (in the case of stock price, it is risk-free interest rate), σ is the volatility, T is time to expiration, and ΔT is the time interval between two consecutive stages.

Option value at each node is the maximum of the two possible choices: value for immediate exercise and value for holding for another period. Applying valuation at each node by working back from the last to the first period, and we can get the value of the option. Table 8-2 presents a binomial tree.

This book proposes a model based on the scenarios established by a binomial tree lattice. In essence, it proposes a new way to look at the binomial tree, recasting it in the form of a stochastic mixed-integer programming model. The idea is to:

Maximize: Expected value on the first node of binomial tree

Subject to: constraints consisting of 0-1 integer variables representing the exercise of the options on each node (= 0 if not exercised, =1 if exercised)

Stochastic mixed-integer programming and real options constraints

The stochastic mixed-integer programming assumes that the economic uncertain parameters in E, e, β_j, or c_j in objective function (Equation 8-1) and constraints (Equation 8-2) and (Equation 8-3) evolve as discrete time stochastic processes with a finite probability space. A scenario tree is used to represent the evolution of an uncertain parameter [Ahmed, King, and Parija, 2003]. Figure 8-3 illustrates the notation. The nodes k in all time stages i constitute the states of the world. δ_i denotes the set of nodes corresponding to time stage i. The path from the **root node** 0 at the first stage to a node k is denoted by $P(k)$. Any node k in the last stage n is a **terminal node**. The path $P(k)$ to a terminal node represents a **scenario**, a realization over all periods from 1 to the last stage *n*. The number of terminal nodes Q corresponds to all Q scenarios. Note there is no recombination structure in this tree representation (each node except the root has a unique parent node). For example, we will break a binomial tree structure as in Figure 8-4, where S is the value of the underlying asset, *u* is up factor, and *d* is the down factor.

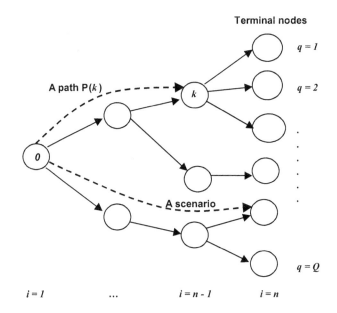

Figure 8-3: Scenario tree

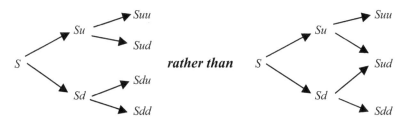

Figure 8-4: Breaking path-independence of a binomial tree

A joint realization of the problem parameters corresponding to scenario q is denoted by

$$\omega^q = \begin{pmatrix} \omega_1^q \\ \cdots \\ \omega_T^q \end{pmatrix},$$

where ω_i^q is the vector consisting of all the uncertain parameters for time stage i in scenario q. p^q denotes the probability for a scenario q. The real options decision variables corresponding to scenario q is denoted by

$$\mathbf{R}^q = \begin{pmatrix} R_1^q \\ \cdots \\ R_T^q \end{pmatrix},$$

where R_i^q is the decision on the option at time stage i in scenario q. 0 denotes no exercise and 1 denotes exercise.

At any intermediate stage i, the decision maker cannot distinguish between any scenario passing through the same node and proceeding on to different terminal node, because the state can only be distinguished by information available up to time stage. Consequently, the feasible solution R_i^q must satisfy:

$$R_i^{q_1} = R_i^{q_2} \qquad \forall (q_1, q_2) \text{ through node } k, \forall k \in \delta_i, \forall i = 1,\dots, n$$

where q_1 and q_2 represent two different scenarios. These constraints are known as **non-anticipativity constraints**.

To illustrate the use of the above approach, we apply it to a standard financial American put option. The formulation is:

$$\text{Max} \quad \sum_{q} p^q \cdot (\sum_{i=1}^{n} E_i^q \cdot R_i^q \cdot e^{-r \cdot \Delta T \cdot (i-1)}) \qquad\qquad \text{Equation 8-8}$$

$$\text{S.t.} \quad E_i^q = S_i^q - K \qquad\qquad\qquad\qquad\qquad \text{Equation 8-9}$$

$$\sum_{i} R_i^q \leq 1 \qquad \forall q \qquad\qquad\qquad\qquad \text{Equation 8-10}$$

$$R_i^{q_1} = R_i^{q_2} \qquad \forall(q_1, q_2) \text{ through node } k, \forall k \in \delta_i, \forall i = 1,...,n \qquad \text{Equation 8-11}$$

$$R_i^q \in \{0,1\} \qquad \forall i, q \qquad\qquad\qquad\qquad \text{Equation 8-12}$$

where S_i^q is the value of underlying asset at time stage i in scenario q, K is the exercise price.

The objective function Equation 8-8 is the expected value of the option along all scenarios. Equation 8-9 can be any equations that specify the exercise condition. Equation 8-10 makes sure that any option can only be exercised at most once in any scenario. Equation 8-11 are the non-anticipativity constraints. We call Equation 8-10 and Equation 8-11 **real options constraints**.

To illustrate and validate the above formulation, consider an example of an American put option without dividend payment. The variables for this example are current stock price S = $20, exercise price K = $18, risk-free interest rate r = 5% per year, volatility σ = 30%, time interval between consecutive states ΔT = 1 year, and time to maturity T = 3 years. So up factor u = 1.35, down factor d = 0.74, move-up probability p_u = 0.51 and move-down probability p_d = 0.49 according to Equation 8-4 to Equation 8-7. A standard binomial lattice gives the value of the options as $2.20 as in Table 8-2.

Now considering the reformulated problem according to equations Equation 8-8 to Equation 8-12. Solve it using GAMS©, the maximum value of the objective function is also 2.20. The optimal solution of 0-1 variables is shown in Table 8-3, and 1 means exercise. The result exactly correspond to that of the ordinary binomial tree (Table 8-2).

Note there is an exercise in scenarios 7 and 8 that is not at the last time point. This proves the formulation can successfully find out early exercise points and define contingency plans for decision makers.

Table 8-2: Binomial tree for the example American put

	Period 1	Period 2	Period 3	Period 4
Stock Price	20.00	27.00	36.44	49.19
Exercise Value	-2.00	-9.00	-18.44	-31.19
Hold Value	2.20	0.69	0.00	0.00
Option Value	2.20	0.69	0.00	0.00
Exercise or not?	No	No	No	No
Stock Price		14.82	20.00	27.00
Exercise Value		3.18	-2.00	-9.00
Hold Value		4.00	1.48	0.00
Option Value		4.00	1.48	0.00
Exercise or not?		No	No	No
Stock Price			10.98	14.82
Exercise Value			7.02	3.18
Hold Value			6.15	0.00
Option Value			7.02	3.18
Exercise or not?			Yes	Yes
Stock Price				8.13
Exercise Value				9.87
Hold Value				0.00
Option Value				9.87
Exercise or not?				Yes

Table 8-3: Stochastic programming result for the example American put

Scenario	Stock Price Realization					Decision			
	$i = 1$	$i = 2$	$i = 3$	$i = 4$	Probability	$i = 1$	$i = 2$	$i = 3$	$i = 4$
$q = 1$	S	Su	Suu	Suuu	0.132	0	0	0	0
$q = 2$	S	Su	Suu	Suud	0.127	0	0	0	0
$q = 3$	S	Su	Sud	Sudu	0.127	0	0	0	0
$q = 4$	S	Su	Sud	Sudd	0.123	0	0	0	1
$q = 5$	S	Sd	Sdu	Sduu	0.127	0	0	0	0
$q = 6$	S	Sd	Sdu	Sdud	0.123	0	0	0	1
$q = 7$	S	Sd	Sdd	Sddu	0.123	0	0	1	0
$q = 8$	S	Sd	Sdd	Sddd	0.118	0	0	1	0

Formulation for the real options timing model

The stochastic mixed-integer programming reformulation is much more complicated than a simple binomial lattice. It is like using a missile to hit a mosquito to value ordinary financial options. But such reformulations empower analysis of complex path-dependent /interdependent real options "in" projects.

The real options timing model decides the possible sequences of implementation of the project design dependent on the actual evolution of the uncertain future. Technical constraints in the screening model are modified in the real options timing model. Since the screening and simulation models have identified the configuration of design, decision variables in the screening model are treated as parameters now in the options timing model, using their respective value in optimal solution from the screening model.

$\overline{\mathbf{Y}}$ is the set for satisfactory configurations of design parameters obtained by the "options identification" stage,

$$\overline{\mathbf{Y}} = \begin{bmatrix} Y_{11} & \cdots & Y_{1j} \\ \vdots & & \vdots \\ Y_{n1} & \cdots & Y_{nj} \end{bmatrix}$$

A vector $(\overline{Y_{n1}}, \overline{Y_{n2}}, ..., \overline{Y_{nj}})$ corresponding to j design parameters (in the screening model, they are j decision variables) in the nth satisfacroty configuration. To describe the fact that only at most one configuration can be implemented, a vector $\mathbf{z}$ is introduced:

$$\mathbf{z} = \begin{bmatrix} z_1 & z_2 & \cdots & z_n \end{bmatrix}, \ z_n \in \{0,1\}, \text{ and } \sum_n z_n \le 1,$$

So $\mathbf{z}\overline{\mathbf{Y}}$ appropriately represents final configuration.

The real options decision variable corresponding to scenario q is expanded to:

$$\mathbf{R}^q = \begin{bmatrix} R_{11}^q & \cdots & R_{1j}^q \\ \vdots & & \vdots \\ R_{i1}^q & \cdots & R_{ij}^q \end{bmatrix}, \quad R_{ij}^q \in \{0,1\}$$

R_{ij}^q denotes the decision on whether to build the feature according to *jth* design parameter for *ith* time stage in scenario q. The objective function Equation 8-8 corresponding to scenario q is denoted by $f^q(\cdot)$. p^q and $f^q(\cdot)$ are derived from the specific scenario tree based on the appropriate stochastic process for the subject under study. The real options constraints Equation 8-10 to Equation 8-11 are concisely denoted by φ. Most importantly, the objective function is modified to get the expected value along all scenarios.

The real options timing model formulation is as follows.

$$\text{Max} \quad \sum_q p^q \cdot f^q(\mathbf{R}^q, \mathbf{z\overline{Y}}) \qquad \text{Equation 8-13}$$

$$\text{s.t.} \quad \mathbf{T} \begin{bmatrix} (\mathbf{z\overline{Y}})_1 R_{i1}^q \\ \vdots \\ (\mathbf{z\overline{Y}})_j R_{ij}^q \end{bmatrix} \geq \mathbf{t} \text{ and } \mathbf{E} \begin{bmatrix} (\mathbf{z\overline{Y}})_1 R_{i1}^q \\ \vdots \\ (\mathbf{z\overline{Y}})_j R_{ij}^q \end{bmatrix} \geq \mathbf{e} \quad \forall q,i \qquad \text{Equation 8-14}$$

$$\mathbf{R}^q \in \varphi \qquad \text{Equation 8-15}$$

$$\sum_n z_n \leq 1 \qquad \text{Equation 8-16}$$

$$R_{ij}^q \in \{0,1\}, z_n \in \{0,1\} \qquad \forall q,i,j,n \qquad \text{Equation 8-17}$$

where $(\mathbf{z\overline{Y}})_j$ represents *jth* element in the vestor of $\mathbf{z\overline{Y}}$. In short, the formulation has an objective function averaged over all the scenarios, subject to three kinds of constraints:

technical, economic, and real options. By specifying the interdependencies by constraints, we can take into account highly complex relationship among projects.

This formulation has important application in real options "in" projects because the real options constraints can be readily added onto the screening model. This formulation can deal with complex interdependency/path-dependency among options by specifying the interdependency/path-dependency in the technical and economic constraints.

8.4. Application of the framework – illustrated by development of strategies for a series of hydropower dams

This section illustrates the application of the framework by an example on the strategy of a river basin development. The case example concerns the development of a hypothetical river basin involving decisions to build dams and hydropower stations in China. The developmental objective is mainly for hydro-electricity production. Irrigation and other considerations are secondary because the river basin is in a remote and barren place.

8.4.1. Screening model to identify options

The screening model is the first cut of the design space that focuses on the important issues and low fidelity in nature, like looking at the system at a 30,000 feet height for an overview. The screening model may be simplified in a number of ways. In the river basin development case example, we simplified the problem by regarding it as a deterministic problem, taking out the stochasticity of the water flow and electricity price. Another simplification is that we regard all the projects are built together at once and neglecting the fact that the projects have to be built in a sequence over a long period of time, in other words, assuming steady state. With such simplifications, we are able to focus our attention on identifying the most interesting flexibility, and leave the scrutiny of the aspects in the following models and studies. The resulted screening model is a non-linear programming model. The sketch of the model is:

Maximize: Net benefit
Subject to: Water continuity constraints
                   Reservoir storage constraints
                   Hydropower constraints
                   Budget constraints

After the screening model established, some detailed consideration and care should be taken when applying it. What uncertain parameters should be examined? What levels of the uncertain variables should be placed into the screening model? After establishing the screening model, we suggest the following steps to use the screening model systematically to search for the interesting real options:

Step 1. List uncertain variables. These could be exogenous or endogenous. They could be market uncertainty, cost uncertainty, productivity uncertainty, technological uncertainty, etc.
Step 2. Find out the standard deviations or volatilities for the uncertain variables. This can be computed by historical data, implied by experts' estimation[1], or estimated from comparable projects.

---

[1] For example, if experts give their most likely estimation, pessimistic estimation (better than 5% of cases) and optimistic estimation (better than 95% of cases), then since 5% and 95% represent the range of $\pm 1.65\sigma$, we can estimate σ from the experts' opinion.

Step 3. Perform sensitivity analysis on the uncertain variables to pick out the several most important uncertain variables for further analysis. Tornado diagram is a useful tool for such sensitivity analysis.

Step 4. List different levels of the important uncertain variable as inputs for the established screening model to identify where the most interesting real options are.

Step 1

The uncertain variables for the river basin development include the price of electricity, fixed cost of reservoir, variable cost of reservoir, variable cost of power plant, and water flow, etc.

For the purpose of illustration, we pick out three important uncertain variables for further scrutiny: electricity price, fixed cost of reservoir, and variable water flow. This is only for illustration purposes, and real studies should examine more uncertain variables - the most important uncertain variables may be out of people's expectation or intuition.

Step 2

The volatility of electricity price is derived by experts' estimation. The author interviewed two Chinese experts on China energy market to get their pessimistic, most likely, and optimistic estimate of the electricity price for 3 years later. The experts reached the optimistic price estimate of 0.315 RMB/kWh (RMB is Chinese currency) and pessimistic estimation of 0.18 RMB/kWh both with 95% confidence[1], which corresponds to a volatility of 6.96% per year (for details see Wang, 2003, pp.101).

---

[1] Note the optimistic estimate of 0.315 RMB/kWh and the pessimistic estimate of 0.21 RMB/kWh imply a mean other than 0.25 RMB/kWh as of the current electricity price. Human's perception does not conform to mathematical rigor.

The standard deviation of the construction cost was estimated from the standard deviation for cost of megaprojects (Flyvbjerg, et al., 2003, pp. 16). The standard deviation is 39%.

The standard deviations of water flow were calculated from historical data. Note since we consider the project for a life span of 60 years, the annual standard deviation should be translate to 60-year average standard deviation by dividing the annual standard deviation by $\sqrt{60}$ (see Table 8-4). $Q_{in,1,y}$ and $Q_{in,2,y}$ represent upstream inflow for season 1 and 2, respectively, for year y. ΔF_{31y} and ΔF_{32y} represent the incremental inflow between site 3 and site 1 for season 1 and 2, respectively, for year y (there are no other incremental inflows in this river basin).

Table 8-4 Parameters for distribution of water flows (Adjusted for 60-years span)

	Mean	Annual s.d.	60-year s.d
$Q_{in,1,y}$	374	87.7	11.3
$Q_{in,2,y}$	283	45.4	5.9
ΔF_{31y}	389	89.8	11.6
ΔF_{32y}	154	9.7	1.3

Step 3

With the understanding of volatility or standard deviation of the three uncertain parameters, we calculate sensitivity and then draw a tornado diagram regarding the change of net benefit due to 1 standard deviation/volatility change of one of the important uncertain variables, with other uncertain variables kept at the expected value. To do so, we just change one variable a time (with 1 standard deviation or volatility) in the screening model, run the optimization and get the corresponding optimal net benefit to draw the tornado diagram. The resulting tornado diagram is as Figure 8-5. Note the skewness for fixed cost for reservoir and waterflow, that is, the change of net benefit is asymmetric if the uncertain variables change positive and negative one volatility or standard deviation. For the fixed cost for reservoir, if cost is about positive one standard deviation, site 3 is not in the solution, and the loss is limited; while if cost is about negative one standard deviation, site 3 is in the solution, and may generate more electricity revenue. For waterflow, the projects are not going to benefit much from overflow and may have to spill overflow water, while the projects have to take the whole loss if water is less than enough.

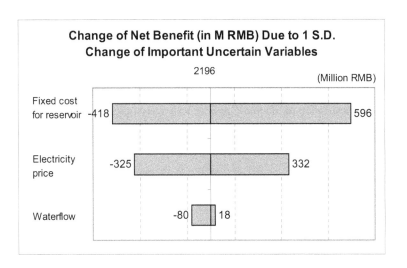

Figure 8-5 Tornado chart for screening model

From the tornado diagram, we understand the uncertainties on fixed cost of reservoir and electricity price are the most important uncertainties. This book studies in-depth only the uncertainty of electricity, however, for the following considerations: This study expands the application of options theory to project design, so the initial interest leans to that related to financial market. Electricity is market traded and its market risk is relatively well understood. Costs are more complicated than electricity price. Electricity price is readily market observable, while the fixed cost has a lot of components and hard to know exactly when the project is being constructed. Moreover, the dynamics of uncertainty of reservoir fixed cost is less understood. Despite the importance of study on uncertainty of fixed cost, the purpose of this book is to lay out a general framework for designing flexibility into projects to deal with uncertainties. So we treat the easier electricity uncertainty as a first step to demonstrate the general framework, and call for further work to study the peculiarity of uncertainty of fixed cost in depth[1].

Step 4

The current electricity price is 0.25 RMB/KWH, but we study the conditions when the electricity price is 0.10, 0.13, 0.16, 0.19, 022, 0.28, 0.31 RMB/KWH. The levels cover most of the range of the experts' pessimistic (0.18 RMB/KWH) to optimistic estimate (0.315 RMB/KMB). To be conservative, we also screen at very low electricity prices as a stress test to see what could happen in the case worse than we would imagine.

Given the 8 levels of electricity price, we get 8 preliminary configurations of the projects. The optimization model is written in GAMS©, and the results are as in Table 8-5. Hs represents the capacity of power plant at site s; Vs represents the capacity of reservoir at site s. "Optimal value" represents the maximum net benefit calculated by the objective function. Note that for case 1, no projects are built; for cases 2 and 3, site 3 is screened out. In a real application considering many more sites, there may be a great number of sites entered the screening models and most of them are screened out. We find that the designs of (H_1 = 3600 MW, V_1 = 9.6 X 10^9 m^3) and (H_2 = 1700 MW, V_2 = 2.5 X 10^7 m^3) are robust with regard to the uncertainty of the electricity price[2]. But for the project at site 3, the optimal design changes when the price of electricity changes, and this is the place on which we should focus designing flexibility.

---

[1] For real application, the analysis should at least be based on both uncertainties of electricity price and fixed cost for reservoir jointly, if there are no other important uncertain variables overlooked.

[2] The only case that these designs are not optimal is when the price of electricity is extremely low - where we do the stress test and is out of the range of experts' estimation.

Table 8-5: Results from the screening model

Case	Electricity Price (RMB/KWH)	H_1 (MW)	V_1 ($10^6 m^3$)	H_2 (MW)	V_2 ($10^6 m^3$)	H_3 (MW)	V_3 ($10^6 m^3$)	Optimal Value (10^6RMB)
1	0.09	0	0	0	0	0	0	0
2	0.12	3600	9600	1700	25	0	0	367
3	0.15	3600	9600	1700	25	0	0	796
4	0.18	3600	9600	1700	25	1564	6593	853
5	0.22	3600	9600	1700	25	1723	9593	1607
6	0.25	3600	9600	1700	25	1946	12242	2196
7	0.28	3600	9600	1700	25	1966	12500	2796
8	0.31	3600	9600	1700	25	1966	12500	3396

Each design for site 1, 2 and 3 represents an option, though the sources of them are subtly different: all options present timing feature, but only option at site 3 has flexible design feature where we consider different reservoir capacity and power plant capacity design. See Table 8-6. Although the features of options have been understood, the final options specification will not be reached until the test of simulation model.

Table 8-6 Sources of options value for designs

	Sources of option value	
	Value of timing	Value of flexible design
Design at site 1	Yes	No
Design at site 2	Yes	No
Design at site 3	Yes	Yes

8.4.2. Simulation model to test other considerations

For the analysis of real options "in" water resources systems, we simulate the combined effect of stochastic variation of hydrologic and economic uncertain parameters. If the time series of the water flow consisted of the seasonal means repeating themselves year after year (no shortages) and the price of electricity were not changing, the simulation model should provide the same results as the screening model. But the natural variability of water flow and electricity price will make the result (net benefit) of each run different, and the average net benefit is not going to be the same as the result from the screening model. The simulated results should be lower. The designs are not going to benefit from excess water when water is more than the reservoir can store. Thus occasional high levels of water do not provide compensation for lost revenues by occasional low levels of water. Due to these uncertainties, the economies of scale apparent under deterministic schemes are reduced.

Satisfying the requirements of various technical considerations such as robustness and reliability, the design with the highest expected benefit from the simulation is those corresponding to the electricity of 0.22 RMB/KWH (Case 5). Note due to the uncertainties in electricity price and water flows, it is not the design corresponding to current electricity of 0.25 RMB/KWH. So the timing options for site 1 and 2 are ($H_1 = 3600$ MW, $V_1 = 9.6$ X 10^9 m^3) and ($H_2 = 1700$ MW, $V_2 = 2.5$ X 10^7 m^3), we have the right to build a project as the specifications, but we do not have the obligation to build them and have the room to observe what happens and decide where to build a project. The option for site 3 contains both timing option and variable design option. We choose 3 design centered Case 5, or ($H_3 = 1564$ MW, $V_3 = 6.93$ X 10^9 m^3), ($H_3 = 1723$ MW, $V_3 = 9.593$ X 10^9 m^3), or ($H_3 = 1946$ MW, $V_3 = 12.242$ X 10^9 m^3). Each design is an option, in that we have the right but not obligation to exercise the option, and the three options are mutually exclusive - only one can be built. A summary for the options sees Table 8-7. This is a portfolio of options. Each option stands for a basic element of flexibility in the project. The next stage will analyze this portfolio of options.

Table 8-7 Portfolio of options for water resources case

	Design specifications	Exercise time
Option at Site 1	H_1 = 3600 MW, V_1 = 9.6 X 10^9 m^3	Any time
Option at Site 2	H_2 = 1700 MW, V_2 = 2.5 X 10^7 m^3	Any time
Option at Site 3	One of $\begin{cases} H_3 = 1564 \text{ MW}, V_3 = 6.93 \text{ X } 10^9 \text{ m}^3 \\ H_3 = 1723 \text{ MW}, V_3 = 9.593 \text{ X } 10^9 \text{ m}^3 \\ H_3 = 1946 \text{ MW}, V_3 = 12.242 \text{ X } 10^9 \text{ m}^3 \end{cases}$	Any time

8.4.3. Real options timing model to analyze real options "in" projects

The general steps to develop and apply a real options timing model (a stochastic mixed-integer programming model) are:

Step 1. Specify real options decision variables, the scenario tree, and the configurations set $\overline{Y}$ for design parameters. The real options decision variables are binary variables corresponding to every option in the portfolio identified. The scenario tree is based the stochastic process that the uncertain variables follow. In the case example, we assume the electricity price follows a Geometric Brownian Motion (GBM) and establish a corresponding scenario tree for the electricity price. This is not necessarily the best model for electricity price: a mean-reverting proportional volatility model might improve the quality of analysis (Bodily and Del Buono, 2002). However, GBM is sufficient to illustrate the analysis framework and the stochastic mixed-integer programming methodology. Configurations set $\overline{Y}$ is as Table 8-8:

Table 8-8 Configurations set $\overline{Y}$ for the river basin case

n	H_1(MW)	V_1(m³)	H_2(MW)	V_2(m³)	H_3(MW)	V_3(m³)
1	3600	9.6 X 10⁹	1700	2.5 X 10⁷	1564	6.93 X 10⁹
2	3600	9.6 X 10⁹	1700	2.5 X 10⁷	1723	9.593 X 10⁹
3	3600	9.6 X 10⁹	1700	2.5 X 10⁷	1946	12.242 X 10⁹

Step 2. Formulation of the stochastic mixed-integer programming. We add real options constraints onto the existing technical and economic constraints, besides the real options constraints stated as Equation 8-10 and Equation 8-11, one constraint needs to force that at most only one design can be built at Site 3, since they are mutually exclusive options. The objective function will also be changed to get the expected value of the scenario tree at the root node. The formulation is as Equation 8-13 to Equation 8-17. The sketch of the formulation is:

Maximize: Expected net benefits
Subject to: Technical and economic constraints
              Real options constraints

Step 3. Establish computer model, find out feasible initial solutions, and run the model.

Results for the case of current electricity price = 0.25 RMB/KWH

Using GAMS©, we obtain the results for the real options timing model as Table 8-9. For example, for the first scenario $q = 1$ that occurs with probability = 0.138: the electricity prices for the first, second, and third 10-year time period (i = 1, 2, and 3) are 0.250, 0.312, and 0.388 RMB/KWH, respectively. The real options decision variables for Project 2 in the first period and Project 1 in the third period are 1's, and the other 7 real options decision variables are 0's (we have 9 real options decision variables for each scenario, 3 projects times 3 periods each). Therefore, for scenario 1, the decision is to build Project 2 in the first period and Project 1 in the third period. The rest of Table 8-9 can be read in the same way. In summary, as in Figure 8-6 the optimal strategy or contingency plan is to build Project 2 in the first time stage. And build nothing in the second stage. In the last stage, we only build Project 1 in the case that price is up for the second stage and up again for the third stage, for other cases, we build nothing. Since Project 3 is not in the solution, the configuration index n can be any of the three as in Table 8-8.

The overall expected net benefit is 4345 million RMB. As a comparison, the timing model without real options considerations (no scenario tree and assume electricity price does not change) suggests that only project 2 should be built in the first time period and that projects 1 and 3 should never be built, with an overall expected net benefit of 4239 million RMB. The value of options is the difference between the optimal benefits from the timing model with and without real options considerations, 106 million RMB. This value is small because of the low electricity of 0.25 RMB/kWh that is also high enough to introduce Project 3 in the contingency plan. Note the valuation of real options "in" projects looks not for an exact numeric result as valuation of financial options, but assesses whether flexible designs are worth. This process about real options valuation is more about the process of designing flexibility itself rather than a specific value of optimal benefit.

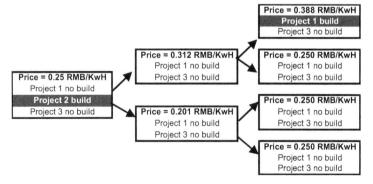

Figure 8-6: Contingency plan for current electricity price of 0.25 KWH/RMB

Table 8-9: Results for real options "in" projects timing model (Current electricity price 0.25 RMB/KWH)

Scenario	Electricity Price Realization				Decision			
	$i = 1$	$i = 2$	$i = 3$	Prob		$i = 1$	$i = 2$	$i = 3$
$q = 1$	0.250	0.312	0.388	0.138	Project 1	0	0	1
					Project 2	1	0	0
					Project 3	0	0	0
$q = 2$	0.250	0.312	0.250	0.233	Project 1	0	0	0
					Project 2	1	0	0
					Project 3	0	0	0
$q = 3$	0.250	0.201	0.250	0.233	Project 1	0	0	0
					Project 2	1	0	0
					Project 3	0	0	0
$q = 4$	0.250	0.201	0.161	0.395	Project 1	0	0	0
					Project 2	1	0	0
					Project 3	0	0	0

Results for the case of current electricity price = 0.30 RMB/KWH

Note when the current price of electricity is 0.25 RMB/KWH as the case example suggests, Project 3 does not enter the solution. In order to test our formulation to see if it is capable of determining best choice among multiple design options for one project, we run the model again at the electricity price of 0.30 RMB/KWH. Now Project 3 enters the solution, and the computer chooses configuration index $n = 3$ with ($H_1 = 3600$ MW, $V_1 = 9.6 \times 10^9$ m^3, $H_2 = 1700$ MW, $V_2 = 2.5 \times 10^7$ m^3, $H_3 = 1564$ MW, $V_3 = 6.93 \times 10^9$ m^3). The resulted contingency plan is as Figure 8-7.

The optimal benefit raised to 7129 million RMB, compared with the result of the timing model without real options considerations, 5195 Million RMB, the options value is 1934 Million RMB, or 37.2% of the result of the the timing model without real options considerations. This option value is much more significant than that in the case of current electricity price of 0.25 RMB/KWh.

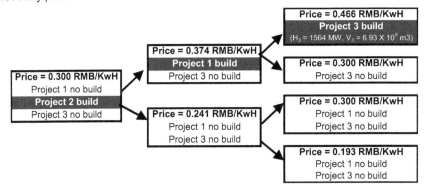

Figure 8-7 Contingency plan for current electricity price of 0.30 KWH/RMB

8.5. Generalizability of framework

The framework proposed in this book is generally applicable to designing flexibility (real options) into other projects. Here we show the application of the framework to another case example on a satellite communications system. The case example builds on the analysis of a satellite system similar to that of Iridium system (de Weck et al., 2004). In 1991, forecasts for the satellite cellular phone market expected up to 3 million subscribers by the year 2000. Initiatives like Iridium and Globalstar were encouraged by the absence of common terrestrial cellular phone standards and slow development of cellular networks at that time. Iridium was designed according to the forecast of 3 million subscribers. However, the rapid success of terrestrial cellular networks and the inconvenient features and high costs of satellite cellular phones soon appeared to doom the two ventures. Iridium only aroused the interest of 50K initial subscribers and filed for bankruptcy in August 1999. Globalstar went bankrupt on February 2002.

Using the framework developed, we analyze the real options of staged deployment of satellites for a system similar to that of Iridium system.

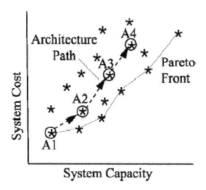

Figure 8-8 Example of a path in trade space [Source: de Weck et al. 2004]

Screening model for satellite system case

Chaize (2003) and de Weck et al. (2004) developed a design space for the satellite communications system by optimization and numerical experiment. The design space is a enumeration of points of system cost and system capacity given various designs based on selection of parameters of orbital altitude, minimum elevation angle, transmit power, antenna diameter, an the use of inter satellite links. After plotting around 1,500 such points, the Pareto Frontier for design is reached. Based on the demand growth scenario, possible staged development path can be recognized. Note that a path is not on the Pareto frontier any more (refer to Figure 8-8), because the staged design sacrifice some benefits of economies of scale, though staged development may prove better in an uncertain environment as shown in the next section. The way to decide each design stage is to decide the capacity requirement for each stage and find the point closest to the Pareto frontier. The major flexibility in the satellite case is repositioning satellite to lower orbit and launching additional satellite to increase the capacity of the system. This model and process is a screening model in effect.

Options analysis for satellite system case

Using the real options – each deployment stage, A1 through A5 – identified by de Weck el al (2004), the authors of this book implemented the real options analysis using the stochastic mixed-integer programming formulation. The result is that the staged deployment has a $0.11 Billion smaller expected cost than that of the best traditional architecture, but the best improvement is that it can take advantage of upside potential, while cut downside risks. The plan can serve up 7.8 million subscribers readily, compared to the best traditional architecture that can only serve up to 2.82 million subscribers. Meanwhile, we would first invest $0.25 Billion to test the market and, if the market is not rosy, we will lose $0.25 Billion rather than $2.01 Billion, the required investment in the best traditional architecture. The downside can be significantly cut. In comparison to what happened to Iridium and Globalstar, this way of project design deserves serious attention.

Similar to the water resources case example, we develop a contingency plan for the development of the satellite communications system as Figure 8-9. Note that architecture A1 and A3 is not built separately in the plan. This is because of the considerations of economies of scale. Building a bigger system, the benefit of economies of scale sometimes overweighs the benefit of postponement of construction (option value and time value of money, investment later has a smaller present value).

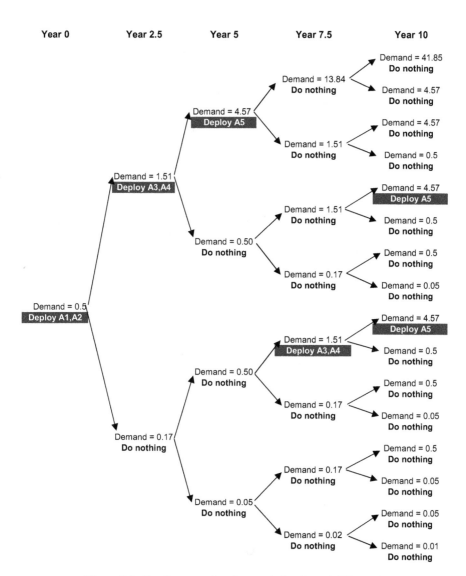

Figure 8-9: Contingency development plan for satellite system

Compared to the work of de Weck et al. (2004), there are two notable differences:

- The saving of expected cost of $ 0.11 B is less than the result in de Weck of $ 0.55 B. This is because this study calculates the expected cost of deployment up to the capacity of 7.8 million users, while de Weck's result was the expected cost of deployment up to the capacity of traditional design, or 2.82 million users.
- This study develops a contingency plan that is not offered by de Weck's study. This contingency plan solves path-dependency problem.

8.6. Discussion

8.6.1. Computational issue

A key consideration in solving a stochastic mixed-integer programming is whether a result is a global or local optimum. It is not simple to prove the result of an mixed-integer programming problem is a global optimum. And it may be hard to find a general solution for the real options timing model because of the special structure of the technical and economic constraints for different projects. Nevertheless, integer programming improves solutions to highly complex and interdependent real options that cannot be solved by ordinary binomial trees. When there is no dependency among nodes, it is possible to optimize on each node and roll back to get the option value. When dependency exists, this simple approach no longer works. A stochastic mixed-integer programming at least provides a local optimum better than the results from conventional approaches or human intuition.

Following is a report on the computational costs: for the river basin development problem, the number of variables is 187, of which 36 are 0-1 discrete variables, and the number of constraints is 261. It takes a laptop (PIII 650, 192M RAM) less than 2 seconds to use GAMS© to figure out a solution. For the satellite case example, the number of variables is 806, of which 400 are 0-1 discrete variables, and the number of constraints is 1131. It takes the laptop (CPU Pentium III 650 MHz, RAM 384M) less than 2 minutes to use GAMS© to figure out a solution.

A natural question is: how efficient is the stochastic mixed-integer programming method? Let us compare it with two alternatives to the stochastic mixed-integer programming method to do options analysis: 1, "brute force" full factorial enumeration method to list all combinations of 0-1 binary decisions, check feasibility of each, discard the infeasible ones, and evaluate remaining set to obtain the optimal combination; 2, the dynamic programming algorithm to analyze options. Several observations on comparison of methods:

1, the algorithm of stochastic mixed-integer programming is highly complex and difficult subject, the solvability of the problem and efficiency of solution is highly dependent on the structure of the problem. Since for every project the technical and economic constraints are different, it is hard to tell if a specific problem can be solved by the stochastic mixed-integer programming efficiently, though the two case examples in this book are solved very smoothly.

2, the framework proposed in this book is a coherent process. The options analysis stochastic mixed-integer programming model adds real options constraints onto the screening model. So once the screening model is developed, it is of minimal effort to extend it into a stochastic mixed-integer programming options analysis model. The computation effort includes modeling effort and machine effort. Since the stochastic mixed-integer programming model is a natural extension of the screening model, the modeling effort is small, though sometimes the machine effort is uncertain.

3, the "brute force" and especially the dynamic programming algorithm serves as good alternatives to the stochastic mixed-integer programming model in case it cannot handle the problem. With more possible algorithm available, the chance of complete solution of the framework is bigger, and the objective of designing flexibility into projects can be better fulfilled.

When the problem becomes extremely large because the combinatorial nature of the problem, we have to limit the stages and get a strategically sound but details less precise analysis.

8.6.2. Insights for different audiences

This research is valuable for the research on engineering systems, real options, and water resources planning. And the method may be thought-provoking for other areas.

Engineering Systems

The framework proposed in this book is **generally applicable** to designing flexibility (real options) into various large-scale engineering systems. It proposes an options identification stage based on optimization and simulation. If only an engineering system of interest can fit the optimization and simulation methodology, the framework developed in this book can be applied. Once screening optimization model and simulation model identify the options, it must be able to develop the real options timing model because the real options timing model is a revision of the screening model with real options constraints added.

This book applies the framework to two examples of engineering systems and demonstrates its applicability. A detailed case example on water resources development can be found in Chapter 6, whose key results are summarized in Section 8.4. Another case example on a satellite communications system as in Chapter 5 also shows the power of the framework.

There may be two major possible complications to apply this framework to a specific engineering system, as we can think of now:

> 1, If the optimization screening model is not suitable for an engineering system? We may still try to figure out another suitable screening model for that system in order to sort out interesting real options.
>
> 2, If there are few robust design parameters from the results of the screening model run - few design parameters proven optimal or near optimal under almost all conditions - and most parameters vary greatly under different conditions? We may have to calculate the sensitivity of the design parameters with regard to the value of the project, pick several most important parameters that affect the value of project greatly, and study carefully the flexibility corresponding to those parameters.

From the results of the case examples on river basin development and satellite communications systems, there is an important insight for engineering systems designers: ***flexibility has great value, which usually means to design smaller and design spare features initially, wait for further information, and prepare for different kinds of happenings.*** The traditional practice is to predict the value of important parameters such as future price and demand and optimize over the expected value of those parameters. Such way tends to lead to false economies of scale. Such economies of scale may actually be suboptimal because the environment is uncertain. If the demand is insufficient, a large design may lead to considerable losses.

Another note for engineering systems design: because of the huge size of the problem for designing large-scale engineering systems in an uncertain environment, both in terms of number of parameters and number of uncertain variables, sometimes, a single model providing the optimal solution is very costly, if not prohibitive, to establish. Instead, we can divide the modeling into a series of models and get a satisfactory solution rather than searching for the best plan among all possibilities. As Herbert Simon pointed out: because of the astronomical amount of extrinsic information and human's limited intrinsic information process capacity, people have only "bounded rationality" and are forced to make decisions not by optimization but by "satisficing". Even though today's computer technologies are tremendously improved, decision problems are still often too big for human beings, especially when high uncertainty is involved. In real decision making, people often do not search for an optimal decision; instead, people stop looking for better decisions once reaching a satisfactory decision. The process - that divides the decision process into several consecutive models and search for a satisfactory solution rather than an optimal solution - is in accordance with the nature of human decision-making in a highly uncertain and complex environment.

Real Options Theory

This research formally attacks the methodology for analyzing real options "in" projects quantitatively. Little work has been done for real options "in" projects. Real options developed from financial options. Most current literature is about real options "on" projects, and based on standard financial theory that depends on the "no arbitrage" and "geometric Brownian motion" assumptions. There are three key difficulties facing the analysis of real options "in" projects, the first is how to deal with options in systems where "no arbitrage" condition is not relevant. The second difficulty: while financial options are well-defined contracts and real options "on" projects are easy to construct via different financial arrangements, it is much harder to identify real options "in" projects where there are myriads of design variables and parameters. The third difficulty: real options "in" projects often exhibit complex path-dependency/interdependency that standard options theory does not dealt with.

This book has successfully dealt with the key difficulties facing real options "in" projects. One of the contributions of this research lies in formally proposing a stage to identify real options "in" projects using screening models, and suggesting the optimization model as one possible screening model to identify real options "in" projects. Another contribution is the new algorithm to calculate options value by stochastic mixed-integer programming and real options constraints. Although such algorithm seems too complicated and costly for usual financial options, it proves a highly useful tool for real options "in" projects where special interdependency/path-dependency is rampant and strategy is a more important issue than a precise valuation.

We hope this study can significantly help other explorers in the new area of real options "in" projects.

Water Resources Planning

This research is valuable for water resources planners in stressing the uncertain nature of the world and pointing out the flexibility that can be employed to manage uncertainties. The change of economic variables not only affects the value of water resources projects but also significantly affects the ability of the projects to finance themselves and repay loans successfully. The real options framework carefully accounts for these uncertain aspects, and increases the economic feasibility of projects and the possibility of ultimate success of projects.

The framework proposed is built on the standard approach for water resources planning, and there is no significant additional cost needed to apply the framework. No technical considerations are lost or overweighed by the real options analysis - real options analysis is only a useful addition on the basis of technically sound designs. The options identification stage needs a number of extra runs of screening and simulation models, but does not need to change the models themselves. The options analysis stage adds real options constraints onto the standard sequencing model, no need to change anything in the sequencing model itself. If the number of time stages considered is small, as we shown previously, the computational cost for the real options timing model is well affordable. In short, once the standard models for water resources planning are established, not much effort is needed to upgrade the models to real options versions.

8.6.3. Further work

Following are several important follow-on research directions:

- How to improve screening models and define a systematic way to identify parameters to vary in screening models;
- How to apply the framework to other projects, especially to see how to deal with a project where optimization is not a possible or relevant design method, and figure out an alternative screening model to identify real options;
- Improve understanding of the nature of the stochastic mixed-integer real options model better and develop better algorithm to solve various problems;
- Better connection between the low fidelity screening model and follow-on high fidelity models;
- It is also interesting to design for a real project with more technical details as well as more economic/social considerations, and study how to describe mathematically the uncertainties of economic/social considerations other than those well studied by finance science; and
- The expansion of the model to base on several important uncertain variables jointly.

8.7. Contributions and Conclusion:

This research develops a comprehensive approach to identify and deal with real options "in" projects, that is, those real options (flexibility) that are integral parts of the technical design. It represents a first attempt to specify analytically the design parameters that provide good opportunities for flexibility for any specific engineering system.

It proposes a two-stage integrated process: options identification followed by options analysis. Options identification includes a screening and a simulation model. Options analysis develops a stochastic mixed-integer programming model to value options. This approach decreases the complexity and size of the models at each stage and thus permits efficient computation even though traditionally fixed design parameters are allowed to vary stochastically.

The options identification stage discovers the design elements most likely to provide worthwhile flexibility. As are often too many possible options for systems designers to consider, they need a way to identify the most valuable options for further consideration, that is, a screening model. This is a simplified, conceptual, low-fidelity model for the system that conceptualizes its most important issues. As it can be easily run many times, it is used to extensively test designs under dynamic conditions for robustness and reliability; and to validate and improve the details of the preliminary design and set of possible options.

The options valuation stage uses stochastic mixed integer programming to analyze how preliminary design identified by the options identification stage should evolve over time as uncertainties get resolved. Complex interdependencies among options are specified in the constraints. This formulation enables designers to analyze complex and problem-specific interdependencies that have been beyond the reach of standard tools for options analysis, to develop explicit plans for the execution of projects according to the contingencies that arise.

The framework developed is generally applicable to engineering systems. The book explores two cases in river basin development and satellite communications. The framework successfully attacks these cases, and shows significant value of real options "in" projects, in the form of increased expected net benefit and/or lowered downside risk.

At the end of this book, the author would point out again that our framework adds significant value to big engineering projects featuring high capital investment - 2 Billion RMB for the river basin example if the electricity price is 0.30 RMB/KWH; and it can help dramatically cut loss if reality is not as good as expected - $0.25 Billion loss instead of $2.01 Billion for the satellite system example! It is my deepest hope that this research can spread, arouse interest, and get tested by practical engineering design!

References

Aberdein, D. A. (1994) "Incorporating risk into power station investment decisions in South Africa," Master of Science Thesis, Sloan School of Management, MIT, Cambridge, MA.

Ahmed, S., King, A.J., and Parija, G. (2003) "A Multi-stage Stochastic Integer Programming Approach for Capacity Expansion under Uncertainty," *Journal of Global Optimization*, 26, pp. 3 – 24.

Alvarado, F. L. and Rajaraman R. (2000) "Understanding price volatility in electricity markets," proceedings of the 33rd Hawaii International Conference on System Sciences, Hawaii.

Amram, M. and Kulatilaka, N. (1999) *Real Options - Managing Strategic Investment in an Uncertain World,* Harvard Business School Press, Boston, MA.

Anderson, E. and Al-Jamal, K. (1995) "Hydraulic-Network Simplification," Journal of Water Resources Planning and Management, 121, pp. 235 – 240.

Archetti, F., DiPillo, G., and Lucertini, M. (1986) *Stochastic Programming,* Springer-Verlag, Berlin, Germany.

Ascher, W. (1978) *Forecasting: An Appraisal for Policy-Makers and Planners,* Johns Hopkins University Press, Baltimore, MD.

Baldwin C.Y. and Clark K.B. (2000) *Design Rules, Vol. 1: The Power of Modularity,* MIT Press, Cambridge, MA.

Baldwin C.Y. and Clark K.B. (2002) "Institutional Forms, Part 1: The Technology of Design and its Problems," Real Options Symposium, University of Maryland, College Park, MD.

Baldwin C.Y. and Clark K.B. (2002) "The Fundamental Theorem of Design Economics," Real Options Symposium, University of Maryland, College Park, MD.

Baldwin, C.Y. and Clark, K.B. (2001) "Modularity after the Crash," HBS working paper, May.

Baxter, M. and Rennie, A. (1996) *Financial Calculus - An Introduction to Derivative Pricing,* Cambridge University Press, Cambridge, UK.

Bellman, R.E. (1957) *Dynamic Programming.* Princeton University Press, Princeton

Bertsimas, D. and Tsitsiklis, J.N. (1997) *Introduction to Linear Programming,* Athena Scientific, Belmont, MA.

Birge, J.R. and Louveaux, F. (1997) *Introduction to Stochastic Programming.* Springer, New York, NY.

Bjerksund, P. and Ekern, S. (1990) "Managing Investment Opportunities Under Price Uncertainty: from 'Last Chance' to 'Wait and See' Strategies," *Financial Management*, Autumn, pp. 65 - 83.

Bodily, S and Del Buono, M. (2002) "Risk and Reward at the speed of Light: A New Electricity Price Model," *Energy Power Risk Management*, Sept., pp. 66 –71.

Boer, F. P. (2002) *The Real Options Solution – Finding Total Value in a High-Risk World,* John Wiley & Sons, New York, NY.

Borison, A. (2003) "Real Options Analysis: Where are the Emperor's Clothes?" 7th Annual Conference on Real Options, Washington DC.
http://www.realoptions.org/papers2003/borison.doc

Brealey, B. A. (1983) An *Introduction to Risk and Return from Common Stocks,* The MIT Press, Cambridge, MA.

Brealey, R. and Myers, S. (2002) *Principles of Corporate Finance*, 6th ed., McGraw-Hill, New York, NY.

Brennan, M. and Schwartz, E. (1985) "A New Approach to Evaluating Natural Resource Investments," *Midland Corporate Finance Journal*, Spring, pp. 37 - 47.

Brach, M.A. (2003) *Real Options in Practice*, John Wiley & Sons, Hoboken, NJ.

Cai, X., McKinney, D., and Lasdon, L. (2003) "Integrated Hydrologic-Agronomic-Economic Model for River Basin Management," *Journal of Water Resources Planning and Management*, Jan – Feb, 129, pp. 4 - 17.

Chaize, M. (2003) "Enhancing the Economics of Satellite Constellations via Staged Deployment and Orbital Reconfiguration," Master of Science Thesis, MIT, Cambridge, MA.

Chenery, H. (1952) "Overcapacity and Acceleration Principle," *Econometrica*, 20, 1, pp. 1-28.

Cherian, J.A. Patel, J., and Khripko, I. (2000) "Optimal Extraction of Nonrenewable Resources When Costs Cumulate," in *Project Flexibility, Agency, and Competition* (2000) pp. 224 - 253, edited by Brennan, M.J. and Trigeorgis, L., Oxford University Press, New York, NY.

Childs, P.D., Riddiough, T.J., and Triantis, A.J. (1996) "Mixed Use and the Redevelopment Option," *Real Estate Economics*, 24, 3, pp. 317 - 339.

Childs, P.D. and Triantis, A.J. (1999) "Dynamic R&D Investment Policies," *Management Science*, 45, 10, pp. 1359 - 1377.

China Development Bank (2002) "Sichuan Yalongjiang River Hydropower Projects Study," Beijing, China.

China State Power (2001) *Guideline for Financial Evaluation*, Beijing, China.

Ciesluk, W. J., Gaffney, L. M., Hulkover, N. D., Klein, L., Olivier, P. A., Pavloff, M. S., Pomponi, R. A., and Welch, W. A. (1992) "An Evaluation of Selected Mobile Satellite Communications Systems and their Environment," MITRE Corporation, ESA contract 92BOOOOO60, April.

Clarke, C. (2000) "Cross-Check Survey," World Commission on Dams Work Program Component, November. http://www.dams.org/kbase/survey/.

Cohon, J.L. and Marks, D.H. (1973) "Chapter 21: Multiobjective Analysis in Water Resource Planning," adapted from *Water Resources Research*, 9, 4, August, pp. 333 – 340. in *Systems Planning and Design* (1974) edited by de Neufville, R. and Marks, D.H. Prentice Hall, Englewood Cliffs, NJ.

Cortazar, G. and Casassus, J. (1997) "A Compound Option Model for Evaluating Multistage Natural Resource Investment," in *Project Flexibility, Agency, and Competition* (2000) pp. 205 - 223, edited by Brennan, M.J. and Trigeorgis, L., Oxford University Press, New York, NY.

Cooper, R.G., Edgett, S.J., and Kleinschmidt, E.J. (1997) "Portfolio Management in New Product Development: Lessons from Leading Firms," Working Paper, Feb, SOB, McMaster University, Ontario, Canada.

Copeland, T.E. and Antikarov, V. (2001) *Real Options - A Practitioner's Guide,* TEXERE, New York, NY.

Copeland T. E. and Keenan, P. T. (1998a) "How much is flexibility worth?" *The McKinsey Quarterly,* 2, pp. 38 - 49.

Copeland, T.E. and Keenan P.T. (1998b) "Making Real Options Real," *The McKinsey Quarterly,* 3, pp. 128 - 141.

Cox, J., Ross, S., and Rubinstein, M. (1979) "Option Pricing: A Simplified Approach," *Journal of Financial Economics,* 7, pp. 263 - 384.

Dahe, P. and Srivastava, D. (2002) "Multireservoir Multiyield Model with Allowable Deficit in Annual Yield," *Journal of Water Resources Planning and Management,* Nov – Dec,128, pp. 406 – 414.

Dantzig, G. (1948), "Programming in a Linear Structure," *Comptroller,* United States Air Force, Washington DC, February.

de Neufville, R. (1990) *Applied Systems Analysis,* McGraw-Hill, New York, NY.

de Neufville, R. (2002) Class notes for Engineering Systems Analysis for Design, MIT engineering school-wide elective, Cambridge, MA.
http://ardent.mit.edu/real_options/Common_course_materials/slides.html

de Neufville, R. and Odoni, A. (2003) *Airport Systems Planning, Design, and Management,* McGraw-Hill, New York, NY.

de Neufville, R., Ramírez, R., and Wang, T. (2005) "Real Options in Infrastructure Design: The Bogotá Water Supply System Case", working paper.

de Neufville, R., Scholtes, S., and Wang, T. (2005) "Valuing Real Options by Spreadsheet: Parking Garage Case Example," *ASCE Journal of Infrastucture Systems,* in press.

de Neufville, R. et al. (2004) "Uncertainty Management for Engineering Systems Planning and Design." Monograph, Engineering Systems Symposium, MIT, Cambridge, MA. March. http://esd.mit.edu/symposium/pdfs/monograph/uncertainty.pdf

de Weck, O., de Neufville, R., and Chaize, M. (2004) "Staged Deployment of Communications Satellite Constellations in Low Earth Orbit," *Journal of Aerospace Computing, Information, and Communication,* March, pp. 119-136.

Dias, M. (2002) "Overview of Real Options in Petroleum," Real Options Seminar Series, Spring, Sloan School of Management, MIT, Cambridge, MA.

Dias, M. "Real Options in Petroleum Investments: Uncertainty, Irreversibility, Timing." at http://www.puc-rio.br/marco.ind/.

Dixit, A.K. and Pindyck, R.S. (1994) *Investment under Uncertainty,* Princeton University Press, Princeton, NJ.

Dixit, A.K. and Pindyck, R.S. (1995) "The Options Approach to Capital Investment," *Harvard Business Review,* May-June, pp.105 - 115.

Draper, A., Jenkins, M., Kirby, K., Lund, J., and Howitt, R. (2003) "Economic-Engineering Optimization for California Water Management," *Journal of Water Resources Planning and Management,* May – June,129, pp. 155 - 164.

Draper, A., Munévar, A., Arora, S., Reyes, E., Parker, N., Chung, F., and Peterson, L. (2004) "CalSim: Generalized Model for Reservoir System Analysis," *Journal of Water Resources Planning and Management,* Nov – Dec, 130, pp. 480 - 489.

Faulker, T. (1996) "Appling 'Options Thinking' to R&D valuation Research," *Technology Management,* May-June, pp. 50 - 56.

Flyvbjerg, B., Bruzelius, N, and Rothengatter, W. (2003) *Megaprojects and Risk – An Anatomy of Ambition*. Cambridge University Press, Cambridge, UK.

Ford, D. N., Lander, D. M. and Voyer, J. J. (2002) "A Real Options Approach to Valuing Strategic Flexibility in Uncertain Construction Projects," *Construction Management and Economics*, 20, pp. 343 – 351.

Fuggle, R. and Smith, W.T. (2000) "Experience with Dams in Water and Energy Resource Development in the People's Republic of China," World Commission on Dams Country Review Paper, November. http://www.dams.org/kbase/studies/cn/.

Gallager, R.G. (1996) *Discrete Stochastic Process*, Kluwer Academic Publishers, Boston, MA.

Geske, R. (1979) "The valuation of compound options," *Journal of Financial Economics*, March, pp. 63 - 81.

Gesner, G. and Jardim, J. (1998) "Bridge within a bridge," *Civil Engineering*, October, pp. 44 – 47.

Grenadier, S. and Weiss, A.M. (1997) "Investment in Technological Innovations: An Options Pricing Approach," *Journal of Financial Economics*, 44, pp. 397 – 416.

Grinblatt, M. and Titman, S. (1998) *Options of Financial Market and Corporate Strategy*. Irwin/McGraw-Hill, Boston, MA.

Hastings, D. (2004) "The Future of Engineering Systems: Development of Engineering Leaders," Monograph, Engineering Systems Symposium, MIT, Cambridge, MA. March. http://esd.mit.edu/symposium/pdfs/monograph/future.pdf

Hlouskova, J., Jeckle, M., and Kossmeier, S. (2002) "Real Options Models and Electricity Portfolio Management," 6th Annual Conference on Real Options, Paphos, Cyprus. http://www.realoptions.org/abstracts/abstracts02.html

Ho, S. P. and Liu, L. Y. (2003) "How to Evaluate and Invest in Emerging A/E/C Technologies under Uncertainty," *ASCE Journal of Construction Engineering and Management,* January/February, pp. 16 – 24.

Howell, S., Stark, A., Newton, D., Paxson, D., Cavus, M., Pereira, J., and Patel, K. (2001) *Real Options – Evaluating Corporate Investment Opportunities in a Dynamic World*, Financial Times Prentice Hall, London, UK.

Hreisson, E.B. (2000) "Economies of Scale and Optimal Selection of Hydroelectric Projects," presented at the IEEE/IEE DRPT2000 Conference, April, City University, London, UK.

Hreisson, E.B. (1990) "Optimal Sizing of Projects in a Hydro-based Power System," *IEEE Transactions on Energy Conversion*, 5, 1, March, pp. 32 - 38.

Huchzermerier, A. and Loch, C.H. (1999) "Project Management Under Risk Using the Real Options Approach to Evaluate Flexibility in R&D," INSEAD Working Paper, March, Cedex, France.

Hufschmidt, M.M. and Fiering, B (1966) *Simulation Techniques for the Design of Water-Resource Systems,* Harvard University Press, Cambridge, MA.

Hull, J.C. (1999) *Options, Futures, and Other Derivatives,* Prentice Hall, Upper Saddle River, NJ.

Hull, J. and White, A. (1993) "Efficient Procedure for Valuing European and American Path-dependent Options," *Journal of Derivatives*, Fall, pp. 21 – 31.

Ito, K. (1951) "On Stochastic Differential Equations," *Memoirs, American Mathematical Society,* 4, pp. 1 - 51.

Jacoby, H.D. and Loucks, D. (1972) "Chapter 17: The Combined Use of Optimization and Simulation Models in River Basin Planning," adapted from *Water Resources Research*, 18, 6, Dec, pp. 1401 – 1414. in *Systems Planning and Design* (1974) edited by de Neufville, R. and Marks, D.H. Prentice-Hall, Englewood Cliffs, NJ.

Johnson, R.A. and Wichern, D.W. (1997) *Business Statistics: Decision Making with Data*, John Wiley & Sons, New York, NY.

Juan, C., Olmos, F., Perez, J.C., and Casasus, T. (2002) "Optimal Investment Management of Harbor Infrastructure – A Options Viewpoint," 6th Annual Conference on Real Options, Paphos, Cyprus. http://www.realoptions.org/papers2002/juan_realoptions.pdf

Kasanen, E. (1993) "Creating Value by Spawning Investment Opportunities," *Financial Management*, Autumn, pp. 251 - 258.

Keefer, D.L and Bodly, S.E. (1983) "Three-Point Approximation for Continuous Random Variables," *Management Science,* 29, 5, pp. 595 - 609.

Keeney, R. and de Neufville, R (1972) "Systems Evolution through Decision Analysis: Mexico City Airport," *Systems Engineering*, 3, 1, 34 - 50.

Kemna, A. G.Z. (1993) "Case Studies on Real Options," *Financial Management*, Autumn, pp. 259 - 270.

Kim. Y and Palmer. R (1997) "Value of Seasonal Flow Forecasts in Bayesian Stochastic Programming," *Journal of Water Resources Planning and Management*, Nov – Dec,123, pp. 327 - 335.

Koekebakker, S. and Sodal, S. (2002) "The Value of an Operating Electricity Production Unit," 6th Annual Conference on Real Options, Paphos, Cyprus. http://www.realoptions.org/papers2002/SodalWP.pdf

Kogut, B. and Kulatilaka, N. (1994) "Operating Flexibility, Global Manufacturing, and the Option Value of a Multinational Network," *Management Science,* 40, 1, pp. 123 - 139.

Kulatilaka, N. (1993) "The Value of Flexibility: The Case of a Dual-Fuel Industrial Steam Boiler," *Financial Management*, Autumn, pp. 271 - 280.

Kulatilaka, N. (1988) "Valuing the Flexibility of Flexible Manufacturing Systems," *IEEE Transactions in Engineering Management*, pp. 250 - 257.

Kumar, R. L. (1995) "An Options View of Investments in Expansion-flexible Manufacturing Systems," *International Journal of Production Economics*, 38. pp. 281 – 291.

Kuhn, H. W. and Tucker, A. W. (1951). "Nonlinear programming," J. Neyman (ed.). *Proceedings of the Second Berkeley Symposium on Mathematical Statistics and Probability*, University of California Press, Berkeley, 481-492.

Lambrecht, B.M. (2002) "The Timing of Takeovers under Uncertainty: A Real options Approach," Real Options Symposium, University of Maryland, College Park, MD.

Laughton, D.G. and Jacoby, H.D. (1993) "Reversion, Timing Options, and Long-term Decision-Making," *Financial Management*, Autumn, pp. 225 - 240.

Lefkoff, L. and Kendall, D. (1996) "Yields from Ground-Water Storage for California State Water Project," *Journal of Water Resources Planning and Management*, Jan – Feb, 122, pp. 72 - 74.

Leslie, K.J. and Michaels M.P. (1997) "The Real Power of Real Options," *The McKinsey Quarterly,* 3, pp. 97 - 108.

Leviakangas, P. L. and Lahesmaa, J. (2002) "Profitability Evaluation og Intelligent Transport System Investments," *ASCE Journal of Transportation Engineering*, May/June, pp. 276 – 286.

Loch, C.H., Pich, M.T., and de Meyer, A. (2000) "Project Uncertainty and Management Styles," INSEAD Working Paper, April, Fontainebleau, France.

Loucks, D., Stedinger, J.R. and Haith, D.A. (1981) *Water Resource Systems Planning and Analysis*, Prentice-Hall, Englewood Cliffs, NJ.

Luehrman, T.A. (1997) "What's it Worth? A General Manager's Guide to Valuation," *Harvard Business Review*, May-Jun, pp. 132 - 142.

Luehrman, T.A. (1998a) "Investment Opportunities as Real Options: Getting Started on the Numbers," *Harvard Business Review*, Jul-Aug, pp. 51 - 67.

Luehrman, T.A. (1998b) "Strategy as a Portfolio of Real Options," *Harvard Business Review*, Sep-Oct, pp. 89 - 99.

Luenberger, D. G. (1998) *Investment Science*, Oxford University Press, New York, NY.

Maass, A., Hufschmidt, M.M., Dorfman, R., Thomas, H.A. Jr., Marglin, S.A., and Fair, G.M. (1962) *Design of Water-Resource Systems*, Harvard University Press, Cambridge, MA.

Major, D.C. and Lenton, L. (1979) *Applied Water Resource Systems Planning*, Prentice-Hall, Englewood Cliffs, NJ.

Major, D.C. (1971) "Chapter 22: Multiobjective Redesign of the Big Walnut Project," adapted from Multiple Objective Redesign of the Big Walnut project by D. C. Major and Associates, Water Resources Council-MIT Cooperative Agreement WRC-69-3 and Multiobjective Resources Planning presented at the Fourth Australian Conference on Hydraulics and Fluid Mechanics, Monash University, Melbourne, Australia, Nov – Dec 1971 by D. C. Major and W. T. O'Brien in *Systems Planning and Design*, edited by de Neufville, R. and Marks, D.H., Prentice-Hall, Englewood Cliffs, NJ.

Major, D.C. and Lenton, R.L. (1979) *Applied Water Resource Systems Planning*, Prentice-Hall, Englewood Cliffs, NJ.

Makridakis, S. (1990) *Forecasting, Planning and Strategy for the 21st Century*, The Free Press, New York, NY.

Makridakis, S., et al. (1984) *The Forecasting Accuracy of Major Time Series Methods*, Wiley, New York, NY.

Makridakis, S. and Hibon, M. (1979a) "Accuracy of Forecasting: An Empirical Investigation," *Journal of the Royal Statistical Society*, Series A, 142, 2, pp. 97-145.

Makridakis, S. and Wheelwright, S. (1979b) *Forecasting*, Vol. 12, Studies in Management Science, North-Holland, Amsterdam, Holland.

Makridakis, S. and Wheelwright, S. (1987) *The Handbook of Forecasting: A Manager's Guide*, Wiley-Interscience, New York, NY.

Makridakis, S. and Wheelwright, S. (1989) *Forecasting Methods for Management*, 5th ed., Wiley, New York, NY.

Manne, A. S. (1961) "Capacity Expansion and Probabilistic Growth," *Econometrica*, 29, 4, Oct, pp. 632 - 649.

Manne, A. (1967) *Investments for Capacity Expansion; Size, Location, and Time-Phasing*, Allen & Unwin, London, UK.

Mason, S.P. and Merton, R.C. (1985) "The Role of Contingent Claims Analysis in Corporate Finance," *Recent Advances in Corporate Finance*, pp. 7 –54, Edited by Altman, E. and Subrahmanyam, M. Richard Irwin, Homewood, IL.

Mauboussin, M.J. (1999) *Get Real - Using Real Options in Security Analysis,* Credit Suisse First Boston Corporation, Boston, MA.

McDonald, R. and Siegel, D. (1986) "The Value of Waiting to Invest," *Quarterly Journal of Economics*, November, 10, pp. 707 – 727.

Merton, R. (1973) "Theory of Rational Option Pricing," Bell Journal of Economics and Management Science, Spring, 4, pp 141 – 183.

Micalizzi, A. and Trigeorgis, L. (2001) *"Eli Lilly - Marketing Strategy of Launching A New Drug with Interacting Expansion Option,"* Real Options Group, Nicosia, Cyprus.

MIT ESD Symposium Committee (2002) "ESD Symposium Committee Overview", Cambridge, MA, May.

Miltersen, K.R. (1997) "Valuation of Natural Resource Investments with Stochastic Convenience Yield and Interest Rates," in *Project Flexibility, Agency, and Competition* (2000) pp. 183 - 204, edited by Brennan, M.J. and Trigeorgis, L., Oxford University Press, New York, NY.

Mittal, G. (2004) "Real Options Approach to Capacity Planning Under Uncertainty," M.S. Thesis, MIT Dept. of Civil and Environmental Engineering, Cambridge, MA.

Mock, R., Schwartz, E., and Stangeland, D. (1989) "The valuation of Forestry Resources under Stochastic Prices and Inventories," *Journal of Financial and Quantitative Analysis*, Dec, pp. 473 - 487.

Montgomery, D.C., Peck, E.A., and Vining, G.G. (2001) *Introduction to Linear Regression Analysis,* John Wiley & Sons, New York, NY.

Morimoto, R. and Hope, C. (2001) "An Extended CBA Model of Hydro Projects in Sri Lanka," Research Papers in Management Studies, Judge Institute of Management, University of Cambridge, UK. http://www.jims.cam.ac.uk/research/working_papers/, WP 15/01.

Morimoto, R. and Hope, C. (2002) "An Empirical Application of Probabilistic CBA: Three Case Studies on Dams in Malaysia, Nepal, and Turkey," Research Papers in Management Studies, Judge Institute of Management, University of Cambridge, UK. http://www.jims.cam.ac.uk/research/working_papers/, WP 19/02.

Moese, P. and Kimball, G. (1946) *Methods of Operations Research*, OEG Rpt. 54, Office of the Chief of Naval Operations, Navy Dept., Washington, D.C.

Moese, P. and Kimball, G. (1951) *Methods of Operations Research*, MIT Press, Cambridge, MA.

Moses, J. (2004) "Fundamental Issues in Engineering Systems: A Framing Paper," Monograph, Engineering Systems Symposium, MIT, Cambridge, MA. March. http://esd.mit.edu/symposium/pdfs/monograph/framing.pdf

Mun, J. (2003) *Real Options Analysis: Tools and Techniques for Valuing Strategic Investments and Decisions*, John Wiley & Sons, Hoboken, NJ.

Myers S.C. and Howe, C.D. (1997) "A Life-Cycle Financial Model of Pharmaceutical R&D," Working paper, Sloan School of Management, MIT, Cambridge, MA.

Myers, S.C. (1987) "Finance Theory and Financial Strategy," *Midland Corporate Finance Journal* , Spring, pp. 6 - 13.

Myers, S.C. (1984) "Finance theory and financial strategy," *Interfaces,* 14, Jan-Feb, pp. 126-137.
Myers, S.C. (1972) "A Note on Linear Programming and Capital Budgeting," *The Journal of Finance,* Mar, pp. 89 - 92.
Neely, III, J. E. and de Neufville, R. (2001) "Hybrid Real Options Valuation of Risky Product Development Projects," *International Journal of Technology, Policy and Management* , 1, 1, pp. 29 - 46.
Nichols, N.A. (1994) "Scientific Management at Merck: An Interview with CFO Judy Lewent," *Harvard Business Review,* Jan-Feb, pp. 89 -99.
Nishikawa, T. (1998) "Water-Resources Optimization Model for Santa Barbara, California," *Journal of Water Resources Planning and Management,* Sep – Oct, 124, pp. 252 - 263.
Oriani, R. and Sobrero, M. (2001) "Market Valuation of Firm's Technological Knowledge: A Real Options Perspective," 21st Annual International Conference of Strategic Management Society, San Francisco, CA.
Oueslati, S.K. (1999) "Evaluation of Nested and Parallel Real Options: Case Study of Ford's investment in Fuel Cell Technology," Master of Science Thesis, Technology and Policy Program, MIT, Cambridge, MA.
Pindyck R.S. and Rubinfeld, D.L. (2001) *Microeconomics,* Prentice-Hall, Upper Saddle River, NJ.
Pindyck, R.S. (1993) "Investment of Uncertain Cost," *Journal of Financial Economics,* Aug, 34, pp. 53 – 76.
Quigg, L. (1993) "Empirical Testing of Real options Pricing Models," *Journal of Finance,* June, pp. 621 - 640.
Ramirez, N. (2002) "Valuing Flexibility in Infrastructure Developments: The Bogota Water Supply Expansion Plan," Master of Science Thesis, Technology and Policy Program, MIT, Cambridge, MA.
Rinnooy Kan, A. H. G. (1986) "Stochastic Integer Programming: The Distribution Problem," *Stochastic Programming,* edited by Archetti, F., DiPillo, G., and Lucertini, M., Springer-Verlag, Berlin, Germany.
Rocha, K., Moreira, A., and David, P. (2002) "Investment in Thermopower Generation: A Real Options Approach for the New Brazilian Electrical Power Regulation," 6th Annual Conference on Real Options, Paphos, Cyprus.
http://www.realoptions.org/papers2002/KatiaThermopower.PDF
Rogers, J. (2002) *Strategy, Value and Risk – The Real Options Approach*, Palgrave Macmillan, New York, NY.
Roos, D. (1998) "Engineering Systems" Report to the MIT Engineering Council by a committee chaired by Dan Roos, Feb.
Roos, D. (2004) "Engineering Systems at MIT – the Development of the Engineering Systems Division," Monograph, Engineering Systems Symposium, MIT, Cambridge, MA. March. http://esd.mit.edu/symposium/pdfs/monograph/history.pdf
Rubinstein, M. (1991) "Double Trouble," RISK, Dec 1991 – Jan 1992, pp 53 – 56.
Sagi, J.S. and Seasholes, M.S. (2002) "Firm-Level Momentum: Theory and Evidence," Real Options Symposium, University of Maryland, College Park, MD.
Savage S. (2000) "The Flaw of Averages," *San José Mercury News*, Oct, San José, CA.
http://www.stanford.edu/~savage/faculty/savage/Flaw%20of%20averages.pdf

Schoenberger, C. R. (2000) "Sidelines," *Forbes*, Dec 25, pp. 32-36.

Schwartz, E. (2002) Patents and R&D as Real Options, UCLA working paper, April, Los Angles, CA.

Schwartz, E.S. and Trigeorgis, L. (2001) *Real Options and Investment under Uncertainty,* MIT Press, Cambridge, MA.

Sengupta, J. K. (1972) *Stochastic Programming – Methods and Applications*, North-Holland Publishing Company, Amsterdam, Netherlands.

Sichuan Hydrology and Hydropower Institute (2002) "The Feasibility Report for Development of Hydropower Stations on Yalongjiang River".

Siegel, D.R., Smith, J.L., and Paddock, J.L. (1987) "Valuing Offshore Oil Properties with Option Pricing Models," *Midland Corporate Finance Journal,* Spring, pp. 22 - 30.

Simon, H. (1957) *Models of Man: Social and Rational*, John Wiley and Sons, New York, NY.

Sinha, A. and Bischof, C. (1998) "Application of Automatic Differentiation to Reservoir Design Models," *Journal of Water Resources Planning and Management*, May – Jun,124, pp 162 – 167.

Sinha, A., Rao, B., and Bischof, C. (1999) "Nonlinear Optimization Model for Screening Multipurpose Reservoir Systems," *Journal of Water Resources Planning and Management*, Jul – Aug,125, pp. 229 - 332.

Sinha, A., Rao, B., and Lall, U. (1999) "Yield Model for Screening Multipurpose Reservoir Systems," Journal of Water Resources Planning and Management, Nov - Dec Vol 125, pp 325 - 332.

Smit, H. T.J. and Ankum, L.A. (1993) "A Real Options and Game-Theoretic Approach to Corporate Investment Strategy Under Competition," *Financial Management,* Autumn, pp. 241 - 250.

Smith, J.E. and Nau, R.F. (1995) "Valuing Risky Projects: Option Pricing Theory and Decision Analysis," *Management Science* , 41, 5, pp. 795 - 816.

Smith, K.W. and Triantis, A. (1993) "The Value of Options in Strategic Acquisition," *Real Options in Capital Investment: New Contributions*, edited by Trigeorgis, L. Praeger, New York, NY, pp. 405 - 418.

Srinivasan, K., Neelakantan, T., Narayan, P., and Nagarajukumar, C. (1999) "Mixed-Integer Programming Model for Reservoir Performance Optimization," *Journal of Water Resources Planning and Management*, Sep – Oct,125, pp. 325 - 332.

Titman, S. (1985) "Urban Land Prices under Uncertainty," *American Economic Review*, June, pp. 505 - 514.

Triantis, A. "Real Options Portal,"
http://www.rhsmith.umd.edu/finance/atriantis/RealOptionsportal.html

Trigeorgis, L. and Mason, S.P. (1987) "Valuing Managerial Flexibility," *Midland Corporate Finance Journal* , Spring, pp. 202 - 224.

Trigeorgis, L. (1993a) "The Nature of Options Interactions and the Valuation of Investments with Multiple Options," *Journal of Financial and Quantitative Analysis,* Spring, pp. 1 - 20.

Trigeorgis, L. (1993b) "Real Options and Interactions with Financial Flexibility," *Financial Management*, Autumn, pp. 202 – 224.

Trigeorgis, L. (1996) *Real Options: Managerial Flexibility and Strategy in Resource Allocation,* MIT Press, Cambridge, MA.

Tufano, P., and Moel, A. (2000) "Bidding for the Antamina Mine – Valuation and Incentives in a Real Option Context," pp. 128 – 150. in *Project Flexibility, Agency, and Competition* edited by Brennan, M., Trigeorgis, L., Oxford University Press, Oxford, UK.

US National Research Council (2004) *Adaptive Management for Water Resources Planning*, National Academies Press, Washington, DC.

US Office of Technology Assessment (1982) *Airport and Air Traffic Control Systems*, U.S. Government Printing Office, Washington, DC.

Wang, T. (2003) "Analysis of Real Options in Hydropower Construction Projects – A Case Study in China," Master of Science Thesis, MIT, Cambridge, MA.

Wang, T. and de Neufville, R. (2004) "Building Real Options into Physical Systems with Stochastic Mixed-Integer Programming," 8th Annual Real Options International Conference, Montreal, Canada.
http://www.realoptions.org/papers2004/de_Neufville.pdf

Wang, T. and de Neufville, R. (2005) "Real Options 'in' Projects," 9th Annual Real Options International Conference, Paris, France.

Watkins, D. and McKinney, D. (1997) "Finding Robust Solutions to Water Resources Problems," *Journal of Water Resources Planning and Management*, Jan – Feb,123, pp. 49 - 58.

William, W. (1996) "Integration of Water Resources Planning and Environmental Regulation,"*Journal of Water Resources Planning and Management*, May – Jun, 122, pp. 189 - 196.

Williams, J. (1991) "Real Estate Development as an Option," *Journal of Real Estate Finance and Economics*, June, pp. 91 - 208.

World Energy Council (1999) Hydropower, Energy Info Center.
http://www.worldenergy.org/wec-geis/publications/reports/ser/hydro/hydro.asp

Zhao, T. and Tseng, C. (2003) "Valuing Flexibility in Infrastructure Expansion," *Journal of Infrastructure Systems*, Sep, pp. 89 – 97.

Zhao, T., Sundararajan, S. K., and Tseng, C. (2004) "Highway Development Decision-Making under Uncertainty: A Real Options Approach," *Journal of Infrastructure Systems*, Mar, pp. 23 – 32.

Appendices

Appendix 3A: Ito's Lemma

Ito's lemma was discovered by a mathematician, K. Ito, in 1951. Suppose that the value of a variable x follows an Ito process:

$$dx = a(x,t)dt + b(x,t)dz$$

where dz is a Wiener process. A function G(x,t) follows the process

$$dG = (\frac{\partial G}{\partial x}a + \frac{\partial G}{\partial t} + \frac{1}{2}\frac{\partial^2 G}{\partial x^2}b^2)dt + \frac{\partial G}{\partial x}bdz$$

where the dz is the same Wiener process. Thus, G also follows an Ito process.

Appendix 5A: GAMS code for American Call Options

```
$title integer programming to solve a binomial tree by Tao Wang on 10/19/2003
$offupper
$offlisting
$offsymxref
$offsymlist
option
limrow=0
limcol=0;

Sets
        i      nodes         /1*4/
        j      scenario      /1*4/;
scalar  deltaT  Delta T           /1/
        sigma  Sigma              /0.3/
        r        risk-free rate    /0.05/
        Scur   Current Stock Price /20/
        K      Striking price      /21/;

Variables
        S(i,j) stock price at node ij
        E(i,j) exercise value at node ij
        H(i,j) holding value at node ij
        V(i,j) option value at node ij
        X(i,j) whether the option exercised at node ij
        OptVal
```

```
        p;

Binary variables
        X(i,j);

X.l('4','1') = 1;
X.l('4','2') = 1;
X.l('4','3') = 1;
X.l('4','4') = 1;
S.l(i,j) = 0;
V.l(i,j) = 0;



Equations
        Obj     objective funtion
        Opt     option value on each node
        Exe     exercise value on each node
        Hld11   holding value on node 11
        Hld21   holding value on node 21
        Hld22   holding value on node 22
        Hld31   holding value on node 31
        Hld32   holding value on node 32
        Hld33   holding value on node 33
        Hld4    holding value on the last time point
        Prb     risk neutral probabilty
        Sto     stock price on each node except the initial stock price
        ;


Obj..      OptVal =e= V('1','1');
Hld11..    H('1','1') =e= (V('2','1')*p + V('2','2')*(1-p))/(exp(r*deltaT));
Hld21..    H('2','1') =e= (V('3','1')*p + V('3','2')*(1-p))/(exp(r*deltaT));
```

```
Hld22..      H('2','2') =e= (V('3','2')*p + V('3','3')*(1-p))/(exp(r*deltaT));
Hld31..      H('3','1') =e= (V('4','1')*p + V('4','2')*(1-p))/(exp(r*deltaT));
Hld32..      H('3','2') =e= (V('4','2')*p + V('4','3')*(1-p))/(exp(r*deltaT));
Hld33..      H('3','3') =e= (V('4','3')*p + V('4','4')*(1-p))/(exp(r*deltaT));
Opt(i,j)..   V(i,j) =e= E(i,j)*X(i,j) + H(i,j)*(1 - X(i,j));
Exe(i,j)..   S(i,j) - K =e= E(i,j);
Hld4(j)..    H('4',j) =e= 0;
Prb..        p =e= (exp(r*deltaT) - exp(-sigma*(deltaT**0.5)))/(exp(sigma*(deltaT**0.5)) -
exp(-sigma*(deltaT**0.5)));
Sto(i,j)..   S(i,j) =e= Scur*exp((ord(i)+1-2*ord(j))*sigma*(deltaT**0.5))


Model tamade /all/;
solve tamade using minlp maximizing OptVal;
display V.l, X.l;
```

Appendix 5B: GAMS Code for American Put Options

```
$title integer programming to solve a binomial tree by Tao Wang on 10/19/2003
$offupper
$offlisting
$offsymxref
$offsymlist
option
limrow=0
limcol=0;

Sets
        m     stages         /1*4/
        s     scenario       /1*8/;
scalar  deltaT Delta T                 /1/
        sigma Sigma                    /0.3/
        r     risk-free rate           /0.05/
        Scur  Current Stock Price      /20/
        K     Striking price           /18/;
Table   St(s,m)  stock prices in scenarios
            1      2      3      4
        1   20    26.997 36.442 49.192
        2   20    26.997 36.442 26.997
        3   20    26.997 20     26.997
        4   20    26.997 20     14.816
        5   20    14.816 20     26.997
        6   20    14.816 20     14.816
```

```
7     20    14.816 10.976 14.816
8     20    14.816 10.976 8.131   ;
Parameters    P(s)   probability associated with scenarios
    /1    0.132
    2     0.127
    3     0.127
    4     0.123
    5     0.127
    6     0.123
    7     0.123
    8     0.118  /;


Variables
    E(s,m)  exercise value
    X(s,m)  whether the option exercised
    OptVal
    con;


Binary variables
    X(s,m);


X.l('8','4') = 0;


Equations
    Obj    objective funtion
    Exe    exercise value on each node
    Opt    option can only be exercises once
    n1     non-antipativity constraint 1
    n2     non-antipativity constraint 2
    n3     non-antipativity constraint 3
    n4     non-antipativity constraint 4
```

```
n5     non-antipativity constraint 5
n6     non-antipativity constraint 6
n7     non-antipativity constraint 7
n8     non-antipativity constraint 8
n9     non-antipativity constraint 9
n10    non-antipativity constraint 10
n11    non-antipativity constraint 11
;


Obj..       OptVal =e= sum(s, sum(m, P(s)*E(s,m)*X(s,m)/exp(r*deltaT*(ord(m)-1))));
Exe(s,m)..   E(s,m) =e= K - St(s,m);
Opt(s)..     sum(m, X(s,m)) =l= 1;
n1(s)..      X(s, '1') =e= con;
n2..        X('1','2') =e=  X('2','2');
n3..        X('2','2') =e=  X('3','2');
n4..        X('3','2') =e=  X('4','2');
n5..        X('5','2') =e=  X('6','2');
n6..        X('6','2') =e=  X('7','2');
n7..        X('7','2') =e=  X('8','2');
n8..        X('1','3') =e=  X('2','3');
n9..        X('3','3') =e=  X('4','3');
n10..       X('5','3') =e=  X('6','3');
n11..       X('7','3') =e=  X('8','3');


Model tamade /all/;
solve tamade using minlp maximizing OptVal;
```

Appendix 5C: GAMS Code for analysis of a

satellite communications system

$title satellite constellation problem by Tao Wang on 03/06/2005

$offupper

$offlisting

$offsymxref

$offsymlist

option

limrow=0

limcol=0;

Sets

       q   scenario   /1*16/

       i   stages   /1*5/

       s   architecture   /1*5/;

alias(i,j);

scalar dr   discount rate   /0.1/

       a   economies of scale factore   /0.03/;

Table   D(q,i) stock prices in scenarios

	1	2	3	4	5
1	0.50	1.51	4.57	7.5	7.5

2	0.50	1.51	4.57	7.5	4.57
3	0.50	1.51	4.57	1.51	4.57
4	0.50	1.51	4.57	1.51	0.50
5	0.50	1.51	0.50	1.51	4.57
6	0.50	1.51	0.50	1.51	0.50
7	0.50	1.51	0.50	0.17	0.50
8	0.50	1.51	0.50	0.17	0.05
9	0.50	0.17	0.50	1.51	4.57
10	0.50	0.17	0.50	1.51	0.50
11	0.50	0.17	0.50	0.17	0.50
12	0.50	0.17	0.50	0.17	0.05
13	0.50	0.17	0.05	0.17	0.50
14	0.50	0.17	0.05	0.17	0.05
15	0.50	0.17	0.05	0.02	0.05
16	0.50	0.17	0.05	0.02	0.01

;

Parameters   P(q)   probability associated with scenarios

l

1	0.057
2	0.060
3	0.060
4	0.062
5	0.060
6	0.062
7	0.062

8   0.065

9   0.060

10   0.062

11   0.062

12   0.065

13   0.062

14   0.065

15   0.065

16   0.068

  /;

Parameters   C(s)   Cost for architecture

    /1   0.25

    2   0.15

    3   0.7

    4   0.8

    5   4.9 /;

Parameters   Cap(s) Incremental capacity for architecture

    /1   0.4

    2   0.1

    3   0.8

    4   1.4

    5   5.1/;

Variables

    OptVal

    con(s)

R(q,i,s)

RTD(q,i,s)

P1(q,i)

P2(q,i);

Binary variables

R(q,i,s);

R.l(q,i,s) = 0;

Equations

Obj   objective funtion

capacity

tech1

tech2

tech3

tech4

tech5

tech6

tech7

tech8

tech9

ro

na1

na2

na3

na4

na5

na6

na7

na8

na9

na10

na11

na12

na13

na14

na15

na16

na17

na18

na19

na20

na21

na22

na23

na24

na25

na26

na27

na28

na29

na30

na31

na32

na33

na34

na35

;

Obj..   OptVal =e=   sum(q, P(q)*sum(i, sum(s, C(s)*R(q,i,s)*(1+dr)**(-ord(i)*2.5+2.5)*(1-a*((R(q,i,'1')+R(q,i,'2')+R(q,i,'3')+R(q,i,'4')+R(q,i,'5'))**1.5))))));

capacity(q,i)..   sum(s, Cap(s)*RTD(q,i,s)) =g= D(q,i);

tech1(q,s)..   RTD(q,'1',s) =e= R(q,'1',s);

tech2(q,s)..   RTD(q,'2',s) =e= R(q,'1',s) + R(q,'2',s);

tech3(q,s)..   RTD(q,'3',s) =e= R(q,'1',s) + R(q,'2',s) + R(q,'3',s);

tech4(q,s)..   RTD(q,'4',s) =e= R(q,'1',s) + R(q,'2',s) + R(q,'3',s) + R(q,'4',s);

tech5(q,s)..   RTD(q,'5',s) =e= R(q,'1',s) + R(q,'2',s) + R(q,'3',s) + R(q,'4',s) + R(q,'5',s);

tech6(q,i)..   R(q,i,'2') =l= RTD(q,i,'1');

tech7(q,i)..   R(q,i,'3') =l= RTD(q,i,'2');

tech8(q,i)..   R(q,i,'4') =l= RTD(q,i,'3');

tech9(q,i)..   R(q,i,'5') =l= RTD(q,i,'4');

ro(q,s)..   sum(i, R(q,i,s)) =l= 1;

na1(q,s)..   R(q,'1',s) =e= con(s);

*

na2(s)..   R('1','2',s) =e= R('2','2',s);

na3(s)..   R('2','2',s) =e= R('3','2',s);

na4(s)..   R('3','2',s) =e= R('4','2',s);

na5(s)..   R('4','2',s) =e= R('5','2',s);

na6(s)..   R('5','2',s) =e= R('6','2',s);

na7(s)..   R('6','2',s) =e= R('7','2',s);

na8(s)..   R('7','2',s) =e= R('8','2',s);

na9(s)..   R('9','2',s) =e= R('10','2',s);

na10(s)..   R('10','2',s) =e= R('11','2',s);

na11(s)..   R('11','2',s) =e= R('12','2',s);

na12(s)..   R('12','2',s) =e= R('13','2',s);

na13(s)..   R('13','2',s) =e= R('14','2',s);

na14(s)..   R('14','2',s) =e= R('15','2',s);

na15(s)..   R('15','2',s) =e= R('16','2',s);

*

na16(s)..   R('1','3',s) =e= R('2','3',s);

na17(s)..   R('2','3',s) =e= R('3','3',s);

na18(s)..   R('3','3',s) =e= R('4','3',s);

na19(s)..   R('5','3',s) =e= R('6','3',s);

na20(s)..   R('6','3',s) =e= R('7','3',s);

na21(s)..   R('7','3',s) =e= R('8','3',s);

na22(s)..   R('9','3',s) =e= R('10','3',s);

na23(s)..   R('10','3',s) =e= R('11','3',s);

na24(s)..   R('11','3',s) =e= R('12','3',s);

na25(s)..   R('13','3',s) =e= R('14','3',s);

```
na26(s)..      R('14','3',s) =e= R('15','3',s);

na27(s)..      R('15','3',s) =e= R('16','3',s);
*

na28(s)..      R('1','4',s) =e= R('2','4',s);

na29(s)..      R('3','4',s) =e= R('4','4',s);

na30(s)..      R('5','4',s) =e= R('6','4',s);

na31(s)..      R('7','4',s) =e= R('8','4',s);

na32(s)..      R('9','4',s) =e= R('10','4',s);

na33(s)..      R('11','4',s) =e= R('12','4',s);

na34(s)..      R('13','4',s) =e= R('14','4',s);

na35(s)..      R('15','4',s) =e= R('16','4',s);



Model tamade /all/;

solve tamade using minlp minimizing OptVal;
```

Appendix 6A: GAMS code for the screening model

```
$title screen model by Tao Wang on 10/19/2003
$offupper
$offlisting
$offsymxref
$offsymlist
option
limrow=0
limcol=0;

*The site and season indices are as follow
Sets
        s    /1, 2, 3/
        t    /1, 2/;

*The following defintion of parameters are based on List of Parameters.
*The order and the unit are strictly folloiwng the List of Parameters.
Parameters
        INFL(t) /1    374
                2    283/;
Parameters
        CAPD(s) /1    9600
                2    25
                3    12500/;
Parameters
        CAPP(s) /1    3600
                2    1700
                3    3200/;
```

Table   DeltaF(s,t)

	1	2
1	0	0
2	0	0
3	212	105;

Scalars es   /0.7/;
Scalars Kt   /15.552/;
Scalars ht   /4320/;
Scalars Yst   /0.35/;
Scalars BetaP /0.25/;
Parameters
      FC(s) /1   11.19
          2   0
          3   8.41/;
Parameters
      VC(s) /1   0.000449
          2   0
          3   0.000668/;
Parameters
      Cheta(s)/1   0.000765
          2   0.00185
          3   0.00088/;
Scalars r   /0.086/;
Scalars crf /0.087/;

*The following defintion of variables are based on List of Variable.
*The order and the unit are strictly folloiwng the List of Variables.
Variables
      yr(s)
      Sst(s,t)
      Xst(s,t)

```
        Est(s,t)
        Pst(s,t)
        Ast(s,t)
        H(s)
        V(s)
        AMAX(s)
        AMIN(s)
        B
        C
        NIB;

Positive variable
        Sst(s,t)
        Xst(s,t)
        Est(s,t)
        Pst(s,t)
        Ast(s,t)
        H(s)
        V(s)
        AMAX(s)
        AMIN(s);

Binary variables
        yr(s);

yr.l('1') = 1;
yr.l('2') = 1;
yr.l('3') = 1;


*The following defintion of equation are based on Complete formulation of the
*mixed-integer screening model
```

Equations

    Obj

    Bene

    Cost

    Cont1

    Cont2

    Cont3

    Cont4

    Cont5

    Cont6

    Cont7

    Cont8

    Cont9

    Cont10

    rese1

    rese2

    rese3

    rese4

    rese5

    Hydro1

    Hydro2

    Hydro3

    Hydro4

    Hydro5

    Hydro6

    Inter1

    ;

*costs and benefits are all in the unit of million RMB

Bene..   B =e= sum(s, (sum(t, BetaP*Pst(s,t))))*0.001;

```
Cost..      C =e= crf*(sum( s, (FC(s)*yr(s) + VC(s)*V(s)*yr(s)))*1000);
Obj..       NIB =e= B - C;


Cont1..     Sst('3','2') - Sst('3','1') =e= (INFL('1') - Xst('3','1'))*kt;
Cont2..     Sst('3','1') - Sst('3','2') =e= (INFL('2') - Xst('3','2'))*kt;
Cont3..     Est('1','1') =e= Xst('3','1') + DeltaF('3','1');
Cont4..     Est('1','2') =e= Xst('3','2') + DeltaF('3','2');
Cont5..     Sst('1','2') - Sst('1','1') =e= (Est('1','1') - Xst('1','1'))*kt;
Cont6..     Sst('1','1') - Sst('1','2') =e= (Est('1','2') - Xst('1','2'))*kt;
Cont7..     Est('2','1') =l= Xst('1','1');
Cont8..     Est('2','2') =l= Xst('1','2');
Cont9..     Sst('2','2') - Sst('2','1') =e= (Est('2','1') - Xst('2','1'))*kt;
Cont10..    Sst('2','1') - Sst('2','2') =e= (Est('2','2') - Xst('2','2'))*kt ;
rese1(s,t).. Sst(s,t) - V(s) =l= 0;
rese2(s)..  V(s) - CAPD(s)*yr(s) =l= 0;
rese3(t)..  Sst('1',t) - 0.14*Ast('1',t)**2 =e= 0;
rese4(t)..  Ast('2',t) =e= 280;
rese5(t)..  Sst('3',t) - 0.15*Ast('3',t)**2 =e= 0;
Hydro1(s,t).. Pst(s,t) - 2.73*es*Kt*Xst(s,t)*Ast(s,t) =l= 0;
Hydro2(s,t).. Pst(s,t) - Yst*ht*H(s) =l= 0;
Hydro3(s,t).. AMIN(s) - Ast(s,t) =l= 0;
Hydro4(s,t).. Ast(s,t) - AMAX(s) =l= 0;
Hydro5(s).. AMAX(s) - 2*AMIN(s) =l= 0;
Hydro6(s).. H(s) - CAPP(s)*yr(s) =l= 0;
Inter1..    yr('2') =l= yr('1');



Model tamade /all/;
solve tamade using minlp maximizing NIB;
display yr.l, H.l, V.l, NIB.l;
```

Appendix 6B: GAMS code for the traditional sequencing model

```
$title sequencing model by Tao Wang on 11/29/2003
$offupper
$offlisting
$offsymxref
$offsymlist
option
limrow=0
limcol=0;

*The site and season indices are as follow
Sets
        s     /1, 2, 3/
        t     /1, 2/
        i     /1, 2, 3/;
alias(i,j);

*The following defintion of parameters are based on List of Parameters.
*The order and the unit are strictly folloiwng the List of Parameters.
Parameters
        INFL(t) /1     374
                2     283/;
Table    DeltaF(s,t)
              1     2
        1     0     0
        2     0     0
```

```
    3    389    154;
Scalars  es    /0.7/;
Scalars  Kt    /15.552/;
Scalars  ht    /4320/;
Scalars  Fst   /0.35/;
Scalars  BetaP  /0.21/;
Parameters
    FC(s)  /1    11.19
           2    0
           3    8.41/;
Parameters
    VC(s)  /1    0.000449
           2    0
           3    0.000668/;
Parameters
    HBar(s) /1    3600
            2    1700
            3    1732/;
Parameters
    VBar(s) /1    9600
            2    0
            3    9593/;
Table   ABar(s,t)
            1    2
   1    262    262
   2    280    280
   3    240    253;
Table   YBar(s,t)
            1    2
   1    0    0
   2    0    0
```

```
   3     -63.6  63.6;
Scalars f     /0.226/;
Parameters
      Cheta(s)/1    0.000765
            2    0.00185
            3    0.00088/;
Scalars r     /0.086/;
Parameters
      PV(i) /1     6.532
            2    2.863
            3    1.254/;
Parameters
      PVO(i) /1     0.896
            2    0.943
            3    0.963/;
Parameters
      PVC(i) /1     1
            2    0.438
            3    0.192/;
Scalars crf    /0.087/;


*The following defintion of variables are based on List of Variable.
*The order and the unit are strictly folloiwng the List of Variables.
Variables
      Xsti(s,t,i)
      Psti(s,t,i)
      Rsi(s,i)
      B
      C
      NIB;
```

Positive variable
    Xsti(s,t,i)
    Psti(s,t,i);

Binary variables
    Rsi(s,i);

Rsi.l('1','1') = 0;
Rsi.l('1','2') = 1;
Rsi.l('1','3') = 0;
Rsi.l('2','1') = 1;
Rsi.l('2','2') = 0;
Rsi.l('2','3') = 0;
Rsi.l('3','1') = 0;
Rsi.l('3','2') = 0;
Rsi.l('3','3') = 1;

*The following defintion of equation are based on Complete formulation of the
*mixed-integer sequencing model

Equations
    Obj
    Bene
    Cost
    Cont1
    Cont2
    Cont3
    Cont4
    Cont5
    Cont6
    Constr

Hydro1

Hydro2

Budget

;

*costs and benefits are all in the unit of million RMB

Bene..   B =e= sum(s, (sum(t, sum(i, PV(i)*BetaP*Psti(s,t,i)*(sum(j$(ord(j) lt (ord(i)+1)),Rsi(s,j))-(1-f)*Rsi(s,i))))))*0.001 + sum(s, sum(t, sum(i, PVO(i)*BetaP*Psti(s,t,i)*Rsi(s,i))))*0.001;

Cost..   C =e= 1000*sum(i, sum(s, Rsi(s,i)*PVC(i)*(FC(s) + VC(s)*VBar(s) + Cheta(s)*HBar(s))));

Obj..   NIB =e= B - C;

Cont1(i)..   Xsti('3','1',i) =e= INFL('1') + YBar('3','1')*sum(j$(ord(j) lt (ord(i)+1)),Rsi('3',j));

Cont2(i)..   Xsti('3','2',i) =e= INFL('2') + YBar('3','2')*sum(j$(ord(j) lt (ord(i)+1)),Rsi('3',j));

Cont3(i)..   Xsti('1','1',i) =e= Xsti('3','1',i) + DeltaF('3','1')+ YBar('1','1')*sum(j$(ord(j) lt (ord(i)+1)),Rsi('1',j));

Cont4(i)..   Xsti('1','2',i) =e= Xsti('3','2',i) + DeltaF('3','2')+ YBar('1','2')*sum(j$(ord(j) lt (ord(i)+1)),Rsi('1',j));

Cont5(i)..   Xsti('2','1',i) =l= Xsti('1','1',i);

Cont6(i)..   Xsti('2','2',i) =l= Xsti('1','2',i);

Constr(s)..   sum(i, Rsi(s,i)) =l= 1;

Hydro1(s,t,i)..   Psti(s,t,i)   =l=   2.73*es*Kt*Xsti(s,t,i)*ABar(s,t)*sum(j$(ord(j) lt (ord(i)+1)),Rsi(s,j)) ;

Hydro2(s,t,i).. Psti(s,t,i) - Fst*ht*HBar(s) =l= 0;

Budget(i)..   sum(s, Rsi(s,i)) =l= 1;

Model tamade /all/;

solve tamade using minlp maximizing NIB;

display Rsi.l, NIB.l;

Appendix 6C: GAMS code for the real options timing model

```
$title sequencing model by Tao Wang on 11/29/2003
$offupper
$offlisting
$offsymxref
$offsymlist
option
limrow=0
limcol=0;

*The site and season indices are as follow
Sets
        s     /1, 2, 3/
        t     /1, 2/
        i     /1, 2, 3/
        q     scenarios     /1, 2, 3, 4/;
alias(i,j);

*The following defintion of parameters are based on List of Parameters.
*The order and the unit are strictly folloiwng the List of Parameters.
Parameters
        INFL(t) /1     374
              2     283/;
Table    DeltaF(s,t)
              1     2
        1     0     0
        2     0     0
        3     389   154;
```

```
Scalars  es     /0.7/;
Scalars  Kt     /15.552/;
Scalars  ht     /4320/;
Scalars  Fst    /0.35/;
Table   BetaP(i,q)     electricity prices in scenatios
           1     2     3     4
    1    0.310  0.310  0.310  0.310
    2    0.498  0.498  0.193  0.193
    3    0.801  0.310  0.310  0.120;
Parameters      P(q)   probability associated with scenarios
    /    1      0.178
         2      0.244
         3      0.244
         4      0.335  /;
Parameters
    FC(s)  /1     11.19
           2   0
           3   8.41/;
Parameters
    VC(s)  /1     0.000449
           2   0
           3   0.000668/;
Parameters
    HBar(s) /1    3600
            2    1700
            3    1732/;
Parameters
    VBar(s) /1    9600
            2    0
            3    9593/;
Table   ABar(s,t)
```

```
        1    2
1     262   262
2     280   280
3     240   253;
```

Table   YBar(s,t)

```
        1    2
1     0    0
2     0    0
3    -63.6  63.6;
```

Scalars f   /0.226/;

Parameters

```
    Cheta(s)/1    0.000765
            2    0.00185
            3    0.00088/;
```

Scalars r   /0.086/;

Parameters

```
    PV(i) /1    6.532
          2    2.863
          3    1.254/;
```

Parameters

```
    PVO(i) /1    0.896
           2    0.943
           3    0.963/;
```

Parameters

```
    PVC(i) /1    1
           2    0.438
           3    0.192/;
```

Scalars crf   /0.087/;

*The following defintion of variables are based on List of Variable.

*The order and the unit are strictly folloiwng the List of Variables.

Variables

    Xstiq(s,t,i,q)

    Pstiq(s,t,i,q)

    Rsiq(s,i,q)

    B1

    B2

    C

    NIB

    con(s);

Positive variable

    Xstiq(s,t,i,q)

    Pstiq(s,t,i,q);

Binary variables

    Rsiq(s,i,q);

Rsiq.l('1','2','1') = 1;

Rsiq.l('1','2','3') = 0;

Rsiq.l('1','3','1') = 0;

Rsiq.l('1','3','2') = 0;

Rsiq.l('1','3','3') = 0;

Rsiq.l('1','3','4') = 0;

Rsiq.l('2','1','1') = 1;

Rsiq.l('3','2','1') = 0;

Rsiq.l('3','2','3') = 0;

Rsiq.l('3','3','1') = 1;

Rsiq.l('3','3','2') = 0;

Rsiq.l('3','3','3') = 0;

Rsiq.l('3','3','4') = 0;

*The following defintion of equation are based on Complete formulation of the
*mixed-integer sequencing model

Equations
    Obj
    Bene1
    Bene2
    Cost
    Cont1
    Cont2
    Cont3
    Cont4
    Cont5
    Cont6
    Hydro1
    Hydro2
    Budget
    opt   option can be exercised only once
    n1   non-antipicativity 1
    n2   non-antipicativity 2
    n3   non-antipicativity 3
    force1
    force2
  force3
    force4
*   force5
    ;

*costs and benefits are all in the unit of million RMB
Bene1..   B1 =e= P('1')*sum(s, (sum(t, sum(i,
PV(i)*BetaP(i,'1')*Pstiq(s,t,i,'1')*(sum(j$(ord(j) lt (ord(i)+1)),Rsiq(s,j,'1'))-(1-

f)*Rsiq(s,i,'1'))))))*0.001 + P('2')*sum(s, (sum(t, sum(i,

PV(i)*BetaP(i,'2')*Pstiq(s,t,i,'2')*(sum(j$(ord(j) lt (ord(i)+1)),Rsiq(s,j,'2'))-(1-

f)*Rsiq(s,i,'2'))))))*0.001 + P('3')*sum(s, (sum(t, sum(i,

PV(i)*BetaP(i,'3')*Pstiq(s,t,i,'3')*(sum(j$(ord(j) lt (ord(i)+1)),Rsiq(s,j,'3'))-(1-

f)*Rsiq(s,i,'3'))))))*0.001 + P('4')*sum(s, (sum(t, sum(i,

PV(i)*BetaP(i,'4')*Pstiq(s,t,i,'4')*(sum(j$(ord(j) lt (ord(i)+1)),Rsiq(s,j,'4'))-(1-

f)*Rsiq(s,i,'4'))))))*0.001;

Bene2..   B2 =e= P('1')*sum(s, sum(t, sum(i,

PVO(i)*BetaP(i,'1')*Pstiq(s,t,i,'1')*Rsiq(s,i,'1'))))*0.001 + P('2')*sum(s, sum(t, sum(i,

PVO(i)*BetaP(i,'2')*Pstiq(s,t,i,'2')*Rsiq(s,i,'2'))))*0.001 + P('3')*sum(s, sum(t, sum(i,

PVO(i)*BetaP(i,'3')*Pstiq(s,t,i,'3')*Rsiq(s,i,'3'))))*0.001 + P('4')*sum(s, sum(t, sum(i,

PVO(i)*BetaP(i,'4')*Pstiq(s,t,i,'4')*Rsiq(s,i,'4'))))*0.001;

Cost..   C =e= P('1')*1000*sum(i, sum(s, Rsiq(s,i,'1')*PVC(i)*(FC(s) + VC(s)*VBar(s)

+ Cheta(s)*HBar(s)))) + P('2')*1000*sum(i, sum(s, Rsiq(s,i,'2')*PVC(i)*(FC(s) +

VC(s)*VBar(s) + Cheta(s)*HBar(s))))+ P('3')*1000*sum(i, sum(s,

Rsiq(s,i,'3')*PVC(i)*(FC(s) + VC(s)*VBar(s) + Cheta(s)*HBar(s)))) + P('4')*1000*sum(i,

sum(s, Rsiq(s,i,'4')*PVC(i)*(FC(s) + VC(s)*VBar(s) + Cheta(s)*HBar(s))));

Obj..   NIB =e= B1 + B2 - C;

Cont1(i,q)..   Xstiq('3','1',i,q) =e= INFL('1') + YBar('3','1')*sum(j$(ord(j) lt

(ord(i)+1)),Rsiq('3',j,q));

Cont2(i,q)..   Xstiq('3','2',i,q) =e= INFL('2') + YBar('3','2')*sum(j$(ord(j) lt

(ord(i)+1)),Rsiq('3',j,q));

Cont3(i,q)..   Xstiq('1','1',i,q) =e= Xstiq('3','1',i,q) + DeltaF('3','1')+

YBar('1','1')*sum(j$(ord(j) lt (ord(i)+1)),Rsiq('1',j,q));

Cont4(i,q)..   Xstiq('1','2',i,q) =e= Xstiq('3','2',i,q) + DeltaF('3','2')+

YBar('1','2')*sum(j$(ord(j) lt (ord(i)+1)),Rsiq('1',j,q));

Cont5(i,q)..   Xstiq('2','1',i,q) =l= Xstiq('1','1',i,q);

Cont6(i,q)..   Xstiq('2','2',i,q) =l= Xstiq('1','2',i,q);

Hydro1(s,t,i,q).. Pstiq(s,t,i,q) =l= 2.73*es*Kt*Xstiq(s,t,i,q)*ABar(s,t)*sum(j$(ord(j) lt

(ord(i)+1)),Rsiq(s,j,q)) ;

```
Hydro2(s,t,i,q).. Pstiq(s,t,i,q) - Fst*ht*HBar(s) =l= 0;
Budget(i,q)..    sum(s, Rsiq(s,i,q)) =l= 1;
opt(s,q)..      sum(i, Rsiq(s,i,q)) =l= 1;
n1(s,q)..       Rsiq(s,'1',q) =e= con(s);
n2(s)..         Rsiq(s,'2','1') =e= Rsiq(s,'2','2');
n3(s)..         Rsiq(s,'2','3') =e= Rsiq(s,'2','4');
force1..        Rsiq('1','2','1') =e= 1;
force2..        Rsiq('2','1','1') =e= 1;
force3..        Rsiq('3','3','1') =e= 1;
force4..        Rsiq('3','3','2') =e= 0;
*force5..        Rsiq('1','3','3') =e= 1;


Model tamade /all/;
solve tamade using minlp maximizing NIB;
```

Appendix 6D: GAMS code for real options timing model considering multiple designs

```
$title sequencing model by Tao Wang on 11/29/2003
$offupper
$offlisting
$offsymxref
$offsymlist
option
limrow=0
limcol=0;

*The site and season indices are as follow
Sets
        s    /1, 2, 3/
        t    /1, 2/
        i    /1, 2, 3/
        n    designs    /1, 2, 3/
        q    scenarios   /1, 2, 3, 4/;
alias(i,j);

*The following defintion of parameters are based on List of Parameters.
*The order and the unit are strictly folloiwng the List of Parameters.
Parameters
        INFL(t) /1    374
                2    283/;
Table   DeltaF(s,t)
                1    2
```

```
    1    0    0
    2    0    0
    3    389    154;
Scalars es    /0.7/;
Scalars Kt    /15.552/;
Scalars ht    /4320/;
Scalars Fst    /0.35/;
```

Table BetaP(i,q)   electricity prices in scenatios

	1	2	3	4
1	0.300	0.300	0.300	0.300
2	0.374	0.374	0.241	0.241
3	0.466	0.300	0.300	0.193;

Parameters   P(q)   probability associated with scenarios

```
    /    1    0.147
         2    0.236
         3    0.236
         4    0.380    /;
```

Parameters

```
    FC(s)  /1    11.19
           2    0
           3    8.41/;
```

Parameters

```
    VC(s)  /1    0.000449
           2    0
           3    0.000668/;
```

Table HBar(s, n)

	1	2	3
1	3600	3600	3600
2	1700	1700	1700
3	1723	1946	1966;

Table VBar(s, n)

```
           1    2    3
    1    9600 9600 9600
    2     0    0    0
    3    9593 12242 12500;
Table   ABar(s,t,n)
           1    2    3
   1.1   262  262  262
   1.2   262  262  263
   2.1   280  280  280
   2.2   280  280  280
   3.1   240  274  277
   3.2   253  286  289;
Table   YBar(s,t)
           1    2
    1     0    0
    2     0    0
    3   -63.6 63.6;
Scalars  f    /0.226/;
Parameters
     Cheta(s)/1   0.000765
           2    0.00185
           3    0.00088/;
Scalars  r    /0.086/;
Parameters
     PV(i)  /1   6.532
           2    2.863
           3    1.254/;
Parameters
     PVO(i) /1   0.896
           2    0.943
           3    0.963/;
```

Parameters

    PVC(i) /1   1

           2   0.438

           3   0.192/;

Scalars crf   /0.087/;

*The following defintion of variables are based on List of Variable.

*The order and the unit are strictly folloiwng the List of Variables.

Variables

    Xstiq(s,t,i,q)

    Pstiq(s,t,i,q)

    Rsiq(s,i,q)

    B1

    B2

    C

    NIB

    con(s);

Positive variable

    Xstiq(s,t,i,q)

    Pstiq(s,t,i,q);

Binary variables

    Rsiq(s,i,q)

    Z(n);

Rsiq.l('1','2','1') = 1;

Rsiq.l('1','2','3') = 0;

Rsiq.l('1','3','1') = 0;

Rsiq.l('1','3','2') = 0;

Rsiq.l('1','3','3') = 0;

Rsiq.l('1','3','4') = 0;
Rsiq.l('2','1','1') = 1;
Rsiq.l('3','2','1') = 0;
Rsiq.l('3','2','3') = 0;
Rsiq.l('3','3','1') = 1;
Rsiq.l('3','3','2') = 0;
Rsiq.l('3','3','3') = 0;
Rsiq.l('3','3','4') = 0;

*The following defintion of equation are based on Complete formulation of the
*mixed-integer sequencing model

Equations
    Obj
    Bene1
    Bene2
    Cost
    Cont1
    Cont2
    Cont3
    Cont4
    Cont5
    Cont6
    Hydro1
    Hydro2
    Budget
    opt   option can be exercised only once
    n1   non-antipicativity 1
    n2   non-antipicativity 2
    n3   non-antipicativity 3
    design

```
        force1
        force2
        force3
        force4
*       force5
        ;
```

*costs and benefits are all in the unit of million RMB

Bene1..   B1 =e= P('1')*sum(s, (sum(t, sum(i,
PV(i)*BetaP(i,'1')*Pstiq(s,t,i,'1')*(sum(j$(ord(j) lt (ord(i)+1)),Rsiq(s,j,'1'))-(1-
f)*Rsiq(s,i,'1'))))))*0.001 + P('2')*sum(s, (sum(t, sum(i,
PV(i)*BetaP(i,'2')*Pstiq(s,t,i,'2')*(sum(j$(ord(j) lt (ord(i)+1)),Rsiq(s,j,'2'))-(1-
f)*Rsiq(s,i,'2'))))))*0.001 + P('3')*sum(s, (sum(t, sum(i,
PV(i)*BetaP(i,'3')*Pstiq(s,t,i,'3')*(sum(j$(ord(j) lt (ord(i)+1)),Rsiq(s,j,'3'))-(1-
f)*Rsiq(s,i,'3'))))))*0.001 + P('4')*sum(s, (sum(t, sum(i,
PV(i)*BetaP(i,'4')*Pstiq(s,t,i,'4')*(sum(j$(ord(j) lt (ord(i)+1)),Rsiq(s,j,'4'))-(1-
f)*Rsiq(s,i,'4'))))))*0.001;

Bene2..   B2 =e= P('1')*sum(s, sum(t, sum(i,
PVO(i)*BetaP(i,'1')*Pstiq(s,t,i,'1')*Rsiq(s,i,'1'))))*0.001 + P('2')*sum(s, sum(t, sum(i,
PVO(i)*BetaP(i,'2')*Pstiq(s,t,i,'2')*Rsiq(s,i,'2'))))*0.001 + P('3')*sum(s, sum(t, sum(i,
PVO(i)*BetaP(i,'3')*Pstiq(s,t,i,'3')*Rsiq(s,i,'3'))))*0.001 + P('4')*sum(s, sum(t, sum(i,
PVO(i)*BetaP(i,'4')*Pstiq(s,t,i,'4')*Rsiq(s,i,'4'))))*0.001;

Cost..   C =e= P('1')*1000*sum(i, sum(s, Rsiq(s,i,'1')*PVC(i)*(FC(s) +
VC(s)*sum(n,VBar(s,n)*z(n)) + Cheta(s)*sum(n,HBar(s,n)*z(n))))) + P('2')*1000*sum(i,
sum(s, Rsiq(s,i,'2')*PVC(i)*(FC(s) + VC(s)*sum(n,VBar(s,n)*z(n)) +
Cheta(s)*sum(n,HBar(s,n)*z(n)))))+ P('3')*1000*sum(i, sum(s, Rsiq(s,i,'3')*PVC(i)*(FC(s)
+ VC(s)*sum(n,VBar(s,n)*z(n)) + Cheta(s)*sum(n,HBar(s,n)*z(n))))) + P('4')*1000*sum(i,
sum(s, Rsiq(s,i,'4')*PVC(i)*(FC(s) + VC(s)*sum(n,VBar(s,n)*z(n)) +
Cheta(s)*sum(n,HBar(s,n)*z(n)))));

Obj..   NIB =e= B1 + B2 - C;
```
```

```
Cont1(i,q)..    Xstiq('3','1',i,q) =e= INFL('1') + YBar('3','1')*sum(j$(ord(j) lt
(ord(i)+1)),Rsiq('3',j,q));
Cont2(i,q)..    Xstiq('3','2',i,q) =e= INFL('2') + YBar('3','2')*sum(j$(ord(j) lt
(ord(i)+1)),Rsiq('3',j,q));
Cont3(i,q)..    Xstiq('1','1',i,q) =e= Xstiq('3','1',i,q) + DeltaF('3','1')+
YBar('1','1')*sum(j$(ord(j) lt (ord(i)+1)),Rsiq('1',j,q));
Cont4(i,q)..    Xstiq('1','2',i,q) =e= Xstiq('3','2',i,q) + DeltaF('3','2')+
YBar('1','2')*sum(j$(ord(j) lt (ord(i)+1)),Rsiq('1',j,q));
Cont5(i,q)..    Xstiq('2','1',i,q) =l= Xstiq('1','1',i,q);
Cont6(i,q)..    Xstiq('2','2',i,q) =l= Xstiq('1','2',i,q);
Hydro1(s,t,i,q).. Pstiq(s,t,i,q) =l=
2.73*es*Kt*Xstiq(s,t,i,q)*sum(n,ABar(s,t,n)*z(n))*sum(j$(ord(j) lt (ord(i)+1)),Rsiq(s,j,q)) ;
Hydro2(s,t,i,q).. Pstiq(s,t,i,q) - Fst*ht*sum(n,HBar(s,n)*z(n)) =l= 0;
Budget(i,q)..    sum(s, Rsiq(s,i,q)) =l= 1;
opt(s,q)..       sum(i, Rsiq(s,i,q)) =l= 1;
n1(s,q)..        Rsiq(s,'1',q) =e= con(s);
n2(s)..        Rsiq(s,'2','1') =e= Rsiq(s,'2','2');
n3(s)..        Rsiq(s,'2','3') =e= Rsiq(s,'2','4');
design..        sum(n, z(n)) =e= 1;
force1..        Rsiq('1','2','1') =e= 1;
force2..        Rsiq('2','1','1') =e= 1;
force3..        Rsiq('3','3','1') =e= 1;
force4..        Rsiq('3','3','2') =e= 0;
*force5..        Rsiq('1','3','3') =e= 1;


Model tamade /all/;
solve tamade using minlp maximizing NIB;
```